BURLEIGH DODDS SCIENCE: INSTANT INSIGHTS

NUMBER 73

Phosphorus uptake and use in crops

burleigh dodds
SCIENCE PUBLISHING

Published by Burleigh Dodds Science Publishing Limited
82 High Street, Sawston, Cambridge CB22 3HJ, UK
www.bdspublishing.com

Burleigh Dodds Science Publishing, 1518 Walnut Street, Suite 900, Philadelphia, PA 19102-3406, USA

First published 2023 by Burleigh Dodds Science Publishing Limited
© Burleigh Dodds Science Publishing, 2023. All rights reserved.

British Library Cataloguing in Publication Data
A catalogue record for this book is available from the British Library

ISBN 978-1-80146-608-0 (Print)
ISBN 978-1-80146-609-7 (ePub)

DOI: 10.19103/9781801466097

Typeset by Deanta Global Publishing Services, Dublin, Ireland

Contents

Series list

Title	Series number
Sweetpotato	01
Fusarium in cereals	02
Vertical farming in horticulture	03
Nutraceuticals in fruit and vegetables	04
Climate change, insect pests and invasive species	05
Metabolic disorders in dairy cattle	06
Mastitis in dairy cattle	07
Heat stress in dairy cattle	08
African swine fever	09
Pesticide residues in agriculture	10
Fruit losses and waste	11
Improving crop nutrient use efficiency	12
Antibiotics in poultry production	13
Bone health in poultry	14
Feather-pecking in poultry	15
Environmental impact of livestock production	16
Pre- and probiotics in pig nutrition	17
Improving piglet welfare	18
Crop biofortification	19
Crop rotations	20
Cover crops	21
Plant growth-promoting rhizobacteria	22
Arbuscular mycorrhizal fungi	23
Nematode pests in agriculture	24
Drought-resistant crops	25
Advances in detecting and forecasting crop pests and diseases	26
Mycotoxin detection and control	27
Mite pests in agriculture	28
Supporting cereal production in sub-Saharan Africa	29
Lameness in dairy cattle	30
Infertility/reproductive disorders in dairy cattle	31
Alternatives to antibiotics in pig production	32
Integrated crop-livestock systems	33
Genetic modification of crops	34

Chapter 1

Advances in understanding plant root uptake of phosphorus

Jiayin Pang, The University of Western Australia, Australia; Zhihui Wen, The University of Western Australia, Australia and China Agricultural University, China; Daniel Kidd and Megan H. Ryan, The University of Western Australia, Australia; Rui-Peng Yu, Long Li and Wen-Feng Cong, China Agricultural University, China; Kadambot H. M. Siddique, The University of Western Australia, Australia; and Hans Lambers, The University of Western Australia, Australia and China Agricultural University, China

1 Introduction

Phosphorus (P) is an important structural component of nucleic acids, coenzymes, phospholipids and sugar phosphates (Veneklaas et al., 2012). Deficiency of P is a major constraint to crop production, and therefore, sustaining food production for a growing world population requires a large input of P fertilisers, which are

http://dx.doi.org/10.19103/AS.2020.0075.16

manufactured from non-renewable resources that are expected to diminish dramatically in the next centuries (Fixen and Johnston, 2012). With the decline of rock phosphate, phosphate fertiliser will inevitably become a scarcer and more costly resource (Cordell et al., 2009). Phosphorus resources are mainly restricted to a few countries, with most rock phosphate reserves found in Morocco, China and the United States (Cordell et al., 2009). At the global scale, P-deficient soils comprise almost 30% of the global cropland area; on the other hand, P has also been overused by applying P fertiliser in excess of crop requirements in many other regions around the world, especially in some developed and rapidly developing countries (MacDonald et al., 2011, Vitousek et al., 2009). High P-fertiliser application is often associated with areas of relatively low P-use efficiency (55% of the global cropland area) (MacDonald et al., 2011). Where long-term applications of P fertiliser have occurred, a legacy soil P bank has built-up which is, however, largely unavailable to most crop plants (Menezes-Blackburn et al., 2018). For soils that exceed the critical values for P (usually defined as the soil P level corresponding to 90% or 95% of maximum yield), there is a need to adopt P maintenance practices to improve financial and environmental outcomes (Weaver and Wong, 2011). Syer et al. (2008) suggested that once critical P values are reached, maintenance rates of P can be based on replacing the amount of P removed in harvested products. Improved crop P-acquisition efficiency would allow lower target critical soil P values and savings in P fertiliser (Simpson et al., 2011), as relatively more P is lost through run-off, leaching and/or soil sorption when P-fertiliser applications are high (MacDonald et al., 2011).

Efficient P management in agriculture will be key to maximise crop productivity in the short- and long-term. Such systems must be constructed to reduce fertiliser-input costs, but without compromising yields, or achieving higher yields with the current rates of fertiliser. This could occur through (1) genetic improvement to increase crop capacity to explore soil for soluble P (Bovill et al., 2013); and/or (2) improved plant capacity to access the large pools of sorbed P (Menezes-Blackburn et al., 2018).

The aim of this chapter is to review our current understanding of crop P acquisition, and to promote future research of the best management of farming systems with a large amount of legacy P, and to suggest key future research directions.

2 Root architecture and morphology associated with phosphorus (P) uptake

Plants are able to respond to varying P availability by changing their root architectural and root morphological specialisations (Lambers et al., 2006, Shen et al., 2011, Lynch, 2019). Compared with plants with an optimum P supply, most P-deficient plants often show an increased root mass fraction, lateral root branching, root hair length and density, and specific root length/surface area (Lynch and Brown, 2008, Kidd et al., 2015, Wen et al., 2019). Conversely, the formation of specialised roots such as cluster roots only occurs in a limited number of species (e.g. in most Proteaceae and some Fabaceae) (Shen et al., 2003, Shane et al., 2006, Lambers et al., 2011). More and more architectural and morphological root traits associated with enhanced P acquisition (in the following sections) have been explored systematically among various species and genotypes within a species (Zhu et al., 2010, Brown et al., 2013, Haling et al., 2018, Sun et al., 2018), and the potential mechanisms underlying their differential P-acquisition efficiency have been discussed.

Large intraspecific variation in root architecture and morphology associated with enhanced P acquisition has been found among genotypes/cultivars within various crop/pasture species: for example, maize (*Zea mays*) (Postma and Lynch, 2011), common bean (*Phaseolus vulgaris*) (Strock et al., 2018), barley (*Hordeum vulgare*) (Schneider and Lynch, 2018) and subterranean clover (*Trifolium subterraneum*) (Jeffery et al., 2017b). For instance, maize genotypes with shallow root growth angles through greater production of crown roots (Bayuelo-Jiménez et al., 2011), with greater lateral root density (Zhu and Lynch, 2004, Postma et al., 2014, Jia et al., 2018), and with longer/denser root hairs (Zhu et al., 2010, Miguel et al., 2015) usually exhibit greater P uptake and plant performance in low-P soils. Moreover, maize genotypes with more root cortical aerenchyma formation (Postma and Lynch, 2011), with greater root cortical senescence resulting in reduced root cortical construction costs (Schneider and Lynch, 2018, Galindo-Castaneda et al., 2019) and common bean genotypes with reduced root secondary growth (Strock et al., 2018) all had advantages in root growth, shoot biomass and P acquisition under limiting soil P availability. Overall, these findings highlight that: (1) root traits that increase soil exploration, especially topsoil-foraging capacity (toward topsoil-foraging ideotypes), enhance P capture, because of the low mobility of P in most soils and its greatest bioavailability in the topsoil layers; (2) root traits that reduce the metabolic cost of soil exploration with lower root-construction costs enable resources to be allocated to other plant functions such as new root growth, P-mobilising exudates and mycorrhizas, which may facilitate overall resource capture and plant performance.

Wide interspecific variation in root architecture and morphology related to contrasting P-acquisition strategies has been demonstrated among different species (Hill et al., 2010, Kidd et al., 2015, Waddell et al., 2015, Yang et al., 2015b, Haling et al., 2016, Lyu et al., 2016, Wen et al., 2019). For example, some species of annual pasture legumes show more than three-fold variation in their critical external P requirement (i.e. P required for 90% of maximum yield) (Haling et al., 2016). Phosphorus-efficient species tend to possess relatively long root hairs, high specific root length and relatively high root length density, and consequently a large root hair cylinder volume (a function of root length, root hair length and average root diameter) at low soil P fertility. This increases soil exploration and thus is associated with a lower critical external P requirement (Kidd et al., 2015, Waddell et al., 2015, Yang et al., 2015b, Haling et al., 2016). Moreover, in response to varying P availability, plants also showed diverse responses (magnitudes and directions) in root architectural and morphological traits to maximise their P capture (Haling et al., 2016, Lyu et al., 2016, Wen et al., 2019). For instance, species with thin roots such as wheat (*Triticum aestivum*) and maize show stronger responses in root branching, first-order root length and specific root length in response to decreasing P availability, while species with thick roots such as faba bean (*Vicia faba*) and chickpea (*Cicer arietinum*) display a limited root morphological response (Lyu et al., 2016, Wen et al., 2019). The grass species *Lolium perenne* and *Rytidosperma richardsonii* increased specific root length in response to suboptimal soil P availability, but different strategies were expressed, with *L. perenne* tending to reduce root tissue density, while *R. richardsonii* tended to increase the proportion of fine roots (Waddell et al., 2015). Collectively, these results showed: (1) wide interspecific differences in root architectural and morphological traits that underpin differences in P-acquisition efficiency and critical P requirement; (2) species-specific variation that fine-tune root architectural and morphological trait(s) in response to varying P availability, with a large variation in root morphological plasticity among species.

Plant-available P shows considerable spatial and temporal variability in natural soils, because it is easily sorbed by soil particles and has an extremely low mobility in soil (Hodge, 2004, York et al., 2016). In response to heterogeneous P distribution (patches) in soil, plant roots exhibit varying extents of proliferation (e.g. lateral root numbers, root growth rate and retention time in the patches), and hence will differ in the extent to which they increase P acquisition from patches (Drew, 1975, Hodge, 2004, Li et al., 2014a, Liu et al., 2015). In general, species with thin roots and/or with fast growth rates showed a greater root morphological plasticity to local P supply than species with thick roots and/ or slow growth rates (Fitter, 1994, Li et al., 2014a, Liu et al., 2015, Zhang et al., 2016, Zhang et al., 2019a). This suggests two contrasting resource-acquisition strategies: an acquisitive strategy versus a conservative strategy. However, the mechanisms of P-dependent changes in root morphology in response to local P supply are not fully understood, and thus merit further study.

Although various adjustments in root architecture and morphology play key roles in enhancing P acquisition and plant growth, no single root morphological trait has been identified that universally increases the ability of the root system to explore the soil for P (Yang et al., 2015b, Waddell et al., 2015, Haling et al., 2016). Ideal root traits that enhance P capture often incur trade-offs for the capture of other resources such as nitrate and water (Bishopp and Lynch, 2015, Lynch, 2015, 2019). Therefore we should consider optimised root morphological traits/trait combinations associated with P acquisition under specific environmental conditions.

In summary, plant roots exhibit great variation in root architecture and morphology associated with P acquisition among species and genotypes within a species. Developing and adopting crop genotypes and species with desirable root traits for improved P-acquisition efficiency would reduce P input and dependence on rock phosphate reserves.

3 Root biochemistry associated with P uptake

3.1 Role of carboxylates in P uptake

The exudation of carboxylates (the anion component of organic acids) into the rhizosphere can give some plants a competitive advantage in P-limiting environments (Jones, 1998). Carboxylates interact with the soil which provides a number of ways to enhance P supply to roots. For example, carboxylates have a strong affinity for the sorption sites occupied by P, thus allowing P to move into the soil solution. They also have the capacity for dissolution of strongly bound P (i.e. P associated with oxides and hydroxides of Fe and Al or with Ca) and the mobilisation of organic P, creating a source of P that was previously unavailable to the plant. Moreover, carboxylates also act as a substrate for micro-organisms and can promote the activity of P-solubilising bacteria in the rhizosphere.

The impact of carboxylates on P acquisition is most evident in plant species that occupy severely P-impoverished soils such as members of the Proteaceae (Shane and Lambers, 2005). However, the finding of significant carboxylate exudation in crop species such as lupins (e.g. *Lupinus albus*) and chickpeas has sparked suggestions about utilising carboxylates to develop crop genotypes with a lower external critical P requirement (Wouterlood et al., 2004) or to enhance plant growth in P-sorbing soils. The most widely studied crop species with respect to carboxylate release are related to the formation of cluster roots in *L. albus*. The mechanism of P mobilisation in the rhizosheath of *L. albus* was investigated under controlled conditions (Gardner et al., 1982a,b, 1983a,b, Gardner and Boundy, 1983). Despite having a root system architecture that was considered unfavourable for efficient nutrient capture (i.e. thick lateral roots and no mycorrhizas), this species was able to survive in P-impoverished soils due to the formation of cluster roots, which exude significant quantities of citrate into the rhizosphere with strongly bound inorganic P ($Ca_3(PO_4)_2$, $FePO_4$, $AlPO_4$). By a process of reduction of Fe^{3+}, acidification and dissolution of phosphate complexes and an increased net uptake of P, the plants maintained growth in P-deficient soils which demonstrates that root physiological activities may be more important than root morphological traits for P acquisition in P-impoverished soils (Lyu et al., 2016).

A number of studies show a role for carboxylates in P mobilisation, with a range of crop species studied with regard to P-fertiliser responses in pots and under field conditions using a wide range of P supplies (Brennan and Bolland, 2001, Bolland et al., 1999, 2000, Bolland and Brennan, 2008, Bolland, 1995, 1997). Figure 1, adapted from Bolland (1997), shows that the two lupin species that produce cluster roots (*L. albus* and *L. cosentinii*) perform better in a field experiment at a low rate of P-fertiliser application than the two species that lack cluster roots (*L. angustifolius* and *L. luteus*), even though these also release carboxylates (Hocking and Jeffery, 2004). Yet, in a review, Lynch (2019) concluded that variation in the production of P-mobilising carboxylates is associated with P mobilisation in vitro, but that rigorous analyses have failed to display the benefits of such variation for P acquisition in a range of soils in the field. These rigorous analyses are obviously available from papers by Bolland and co-workers.

There are few field studies that conclusively demonstrate the benefits of P mobilisation by species such as lupin on subsequent crops, because these fields are often managed with relatively high levels of P fertiliser (Angus et al., 2015). Profitable agricultural practices in developed countries rarely operate in the realms of P deficiency where P mobilisation from carboxylates has proven most effective. For example, significant dry-matter improvements from carboxylate exudation in wheat were reported in solution culture by Ryan et al. (2014), but increases attributed to citrate efflux could only be apportioned to two of six pot studies (and only under severe P deficiency), and there was no yield benefit under field conditions. Citrate is a substrate for microbes and it seems that the released citrate was rapidly consumed in the rhizosphere (Martin et al., 2016). A further complication is that there are nutrient deficiencies, other than P, that can drive changes in root exudation. For example, excessive cation uptake has been linked to increases in malate production and proton release by plants (Yan et al., 1992), and rapid rates of nitrate uptake were positively correlated with increased synthesis of organic anions (Haynes, 1990), particularly citrate and malate (Kirkby and Mengel, 1967). Deficiency of molybdenum, iron and manganese may also cause an imbalance in the ratio of cations to anions which can stimulate an increase in carboxylate exudation.

Despite other potential nutritional factors, P deficiency remains the major driver for carboxylate research. There is no doubt that plants that release far more carboxylates than wheat does, in particular those that produce cluster roots, show enhanced P acquisition and growth when soil P is limiting.

3.2 Finding practical applications for root exudate research

Our knowledge of the role and function of root exudates is continuously advancing, yet there remains a distinct lack of sampling techniques for

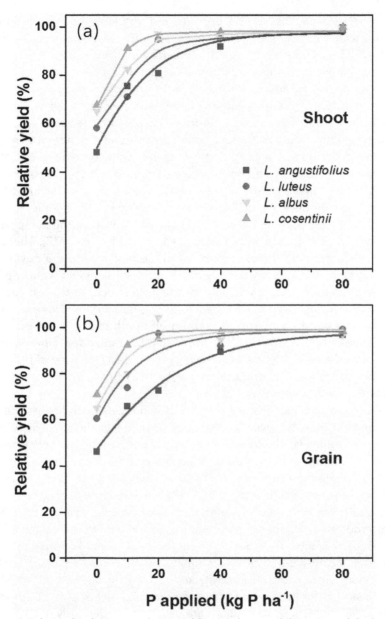

Figure 1 Relationship between relative yield in (a) shoot and (b) grain and the level of phosphorus (P) applied. Lines are fitted using the Mitscherlich equation. The dots and lines in red and blue represent two species without cluster roots (*Lupinus angustifolius*, *L. luteus*), while yellow and green represent two species with cluster roots (*L. albus*, *L. cosentinii*). Adapted from Bolland (1997).

field-grown plants. The relative benefits and shortcomings of the various root exudate collection techniques have been discussed previously (Jones et al., 2003, Oburger et al., 2013, Oburger and Jones, 2018). Generally, growing plants in solution culture alleviates the effects that microbial degradation and soil adsorption can have on plant-derived carboxylates, but results in potentially large sample volumes and precludes the typical root architectural responses to soil. Soil-grown plants often require destructive sampling techniques, which cause root damage and the potential release of carboxylates from root cells, and results vary with soil type (Clark, 1967, Oburger et al., 2009) and soil moisture (Fan et al., 2015). Moreover, the microbial populations in soil not only have the ability to utilise exudates, but can also produce them (Rózycki and Strzelczyk, 1986). Thus, for soil-grown plants, the presence of soil microbes may influence the composition and amount of rhizosphere exudates (Barber and Lynch, 1977). There are few studies of root exudates carried out in the field due to the destructive sampling techniques required. However, rhizosheath sampling is one method that has been used successfully, and despite its destructive nature, the results are surprisingly consistent and repeatable. This method involves excavating plants, removing the bulk soil and rinsing off the rhizosheath (Schefe et al., 2008, Veneklaas et al., 2003, Wang et al., 2017b). It provides information on the root exudates at the time of sampling, and reflects the total of plant and microbial processes. Therefore, field analysis of carboxylates is possible, but results do come with some limitations.

When measuring carboxylates in controlled conditions, careful consideration must be given to the supplied nutrients and pH. For solution culture experiments, the result is a homogenous nutrient concentration where roots and root hairs grow unimpeded, and plants lack the capacity to significantly alter root architecture. The ionic strength of the solution can also have significant impacts (Blamey et al., 1983). While soil studies are a step closer to field-relevant outcomes with regard to carboxylate exudation, the effects of microbes and pathogens can influence root growth, root architecture and hence root exudation. Thus, discussions around soil pasteurisation to minimise their influence must be considered. Often, substrates such as river sand, which have reportedly less microbial activity and fewer adsorption sites, can offer a useful medium for carboxylate measurements (Ryan et al., 2012). However, once added, nutrient concentrations are usually homogenously distributed in the profile, and there is no spatial heterogeneity with regards to soil pH and nutrients, particularly the stratified positioning of P and other nutrients in the soil profile. Repacking the soil to emulate the field soil profile or stratifying soil nutrients significantly alters root architecture (Haling et al., 2016), and hence the distribution of root exudates in the soil profile. The verification of findings from controlled conditions to field experimentation must occur, if results are to have any agricultural or ecological relevance.

There are other abiotic factors that affect carboxylate exudation which are often not controlled in pot studies or are simply too difficult to replicate. For example, temperature is often set at a constant day and night limit, and rarely reflects the daily fluctuations that occur. Equally, light intensity varies seasonally and diurnally and can be highly variable throughout the day due to cloud cover, yet is often set at a constant intensity and tends not to reach the high levels of natural sunlight. Both factors affect photosynthesis and root exudation (Watt and Evans, 1999). Light intensity can also be influenced by shading. While many crop plants often have an inter-row spacing that allows some light penetration to lower levels in the canopy, forage plants are naturally grown in dense swards, where the competitive effects of shading influence plant growth. Stems and petioles become etiolated when exposed to light with a low-red:far-red ratio, and root architecture and root exudation are significantly altered as a result. Experiments that incorporate reflective sleeves around pots that are continually raised to canopy height have been used in forage crop pot trials to emulate field conditions (Jeffery et al., 2017a), and may significantly increase root exudation (>10-fold increase per unit of root length), compared with unconstrained plant growth. Significant advances in controlled-environment technology (i.e. LED lighting) provide the capacity to simulate dawn and dusk lighting to make day/night transitions less abrupt, and enable light intensities approaching that of uninterrupted sunlight. Therefore, some of these abiotic factors may be better simulated in future experiments.

Due to the vast array of sampling techniques used in this field, there are also a number of ways of standardising the data, and this creates difficulties in comparing experimental outcomes (Oburger and Jones, 2018). Although root exudation occurs primarily at the root tips, it is routinely recommended that data are presented as micromoles of root exudate per gram of root dry mass ($\mu mol\ g^{-1}$ root DM). This kind of standardised measurement allows direct comparison with other research, and is very useful when comparing root exudation from members of the same species. However, when comparing a range of species with different root architectural types, this method may not adequately differentiate the species. For example, in Kidd et al. (2018), comparing a species with a comparatively large root diameter and low specific root length (i.e. chickpea) with a number of species with very small root diameters and high-specific root length (i.e. *Lotus* and *Ornithopus* spp.), standardising data by root DM gave distinctly higher values for species with long, thin roots, suggesting their carboxylate exudation was similar to that of chickpea. However, once the root exudation was standardised by root length, the results clearly showed that chickpea had greater concentrations in the

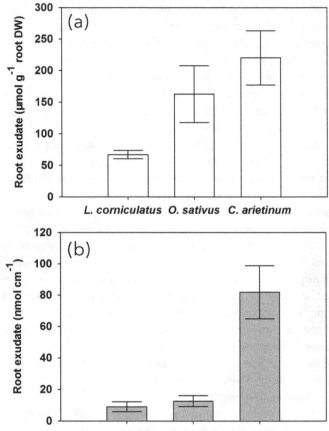

Figure 2 Root exudates collected from the rhizosheath of birdsfoot trefoil (*Lotus corniculatus*), which has a medium specific root length (SRL), French serradella (*Ornithopus sativus*), which has a high SRL, and chickpea (*Cicer arietinum*), which has a low SRL. Amounts of root exudates were not significantly different for chickpea and French serradella when expressed per unit of root dry mass (DM) (a), but they were when root length was used to express the data (b). Adapted from (Kidd et al., 2018).

rhizosheath due to its lower total root length (Fig. 2). This suggests that for comparisons across species, carboxylate amount is best expressed per unit of root length, because these units better reflects differences in root exploration capacity and subsequent interaction between the root and rhizosheath.

It seems that for carboxylate research there is no single method that perfectly addresses the complexity of interactions in field-grown plants. However, if results are repeatable and adequately benchmarked or referenced against previous research, then the use of hydroponics, pot studies or field measurements will be valuable in advancing this research area.

3.3 Role of carboxylates in intercropping systems

Plant species vary widely in their capacity to mobilise sorbed inorganic and organic P (Pearse et al., 2006). We consider plant species that release carboxylates, protons or phosphatases into the rhizosphere, and readily convert unavailable soil P into an available form, P-mobilising species. In contrast, species with a weak or no capacity to access poorly available soil P are non-P-mobilising species (Lambers et al., 2008). Phosphorus-mobilising species may increase the P content and biomass of neighbouring non-P-mobilising species via increased P availability in the rhizosphere, that is, rhizosphere P facilitation (Li et al., 2014b, Waddell et al., 2015, Lambers et al., 2018). The mechanisms underlying interspecific facilitation result in greater P acquisition from both inorganic and organic P sources. Species like faba bean, lupin and *Banksia* spp. release a large amount of protons and carboxylates (e.g. citrate and malate), and thus mobilise sorbed inorganic P (e.g. P bound to oxides and hydroxides of aluminium or iron in acid soils or to calcium in alkaline soils) into plant-available forms that they take up, and this facilitates the growth of their neighbours (Li et al., 2007, Waddell et al., 2015, Dissanayaka et al., 2015, Lambers et al., 2018). Species like chickpea and white lupin that efficiently access organic P because of exudation of phosphatases into the rhizosphere which hydrolyse organic P into inorganic P also benefit the neighbours when growing in close proximity (Li et al., 2004, Li et al., 2003, Waddell et al., 2015, Dissanayaka et al., 2015).

Root barriers are used to measure interspecific rhizosphere effects in glasshouse (Li et al., 2007, Li et al., 2004, Muler et al., 2014, Teste et al., 2014) and field experiments (Li et al., 1999, 2007). A 37-μm nylon mesh barrier blocks root intermingling, but permits exchange of root exudates and hyphal penetration, and a solid barrier eliminates any root interaction by preventing root exudates moving between root zones of two species (Li et al., 2007). If there is greater performance of non-P-mobilising species with a mesh barrier (with root interaction) than with a solid barrier (no root interaction), this provides direct evidence of positive interaction via rhizosphere processes. Interspecific facilitation by P-mobilising species leading to greater P content and performance on P-deficient soils has been observed in several intercropping systems/natural species mixtures in the field and in glasshouse pot experiments: for example, wheat/white lupin (mobilising crop species) (Gardner and Boundy, 1983), wheat/chickpea (mobilising crop species) (Li et al., 2003), maize/faba bean (mobilising crop species) (Li et al., 2007), maize/chickpea (mobilising crop species) (Li et al., 2004), maize/white lupin (mobilising crop species) (Dissanayaka et al., 2015), *Scholtzia involucrata/Banksia attenuata* (mobilising wild species) (Muler et al., 2014) and *Cleistogenes squarrosa/Melissilus ruthenicus* (mobilising wild species) (Yu et al., 2020).

In cropping systems, there have been only a few studies showing the facilitative effects of plants exuding large amounts of carboxylates on neighbouring plants that release little carboxylates. For example, in a maize/faba bean intercropping system under field conditions, the yield of intercropped maize and faba bean was 43% and 26% greater, respectively, than that in monoculture (Li et al., 2007). Another example is intercropping of durum wheat and common bean (Li et al., 2008). More often, intercropping results in a yield improvement of one crop species to the detriment of the companion species as for white lupin with wheat or maize (Li et al., 2010) or faba bean with maize (Li et al., 2010). In some instances, there are yield penalties for both plants, as for intercropped maize and peanut (Inal et al., 2007).

There are still a number of questions related to the benefits of using P-mobilising species for the purposes of P acquisition. Plants grown under P deficiency will often exude greater quantities of carboxylates (Grierson, 1992, Lipton et al., 1987), and for *L. albus*, increasing P supply will lead to a decrease in the formation of cluster roots and subsequent P mobilisation due to citrate efflux. Thus, for a companion species to receive the P-derived benefits from lupin, the soil would need to be relatively P deficient. The ultimate goal of such research is for intercropping of these P-mobilising species with crop species that lack this trait. Through this association, organic P or insoluble inorganic P can be made available to the crop plant through release of carboxylates, protons and phosphatases (Li et al., 2014b). In some cases, the improved yield of intercropping of two species in low-P soil is more likely due to 'complementarity' or 'niche differentiation', with the companion species exploring different soil horizons and/or different soil P sources (Becquer et al., 2017). So, the question of in what kind of agricultural system would this method of P acquisition be functional or even profitable deserves further study.

There are some examples where a low-input agricultural system might benefit from P mobilisation through intercropping. These include areas of sub-Saharan Africa, where P fertilisers are largely unavailable or where phosphate buffering of the soil is so high that the rates of P fertiliser required to obtain a growth response are cost prohibitive. In developed countries, perhaps a more pertinent system lies in a resurgent organic agricultural industry, where there are no synthetic fertiliser inputs at all, and soil organic P inputs are insufficient. However, some of the other functional traits associated with root exudates (including carboxylates) will also be important such as the promotion of beneficial interactions with plant-growth promoting micro-organisms (Wang and Lambers, 2019). The control of rhizosphere micro-organisms by the plant through root exudates is likely the next frontier in rhizosphere research. Therefore, further studies into the production of not only carboxylates, but also other root exudates in various plant species will prove increasingly valuable. The selection of appropriate plant species to match soil chemical properties with plant root exudates that facilitate nutrient acquisition might greatly benefit agricultural production.

3.4 Leaf manganese (Mn) as a proxy of belowground carboxylates for plant breeding

Belowground functional traits, such as exudation of carboxylates, have become increasingly interesting to plant scientists; however, as discussed above, the measurement of carboxylates in the rhizosphere is often laborious (Lambers et al., 2015a). Shane and Lambers (2005) found that the Mn concentration in old leaves of *Hakea prostrata* (a species with cluster roots native to Australia) increased with an increasing percentage of root mass invested in cluster roots, attributed to the increased release of carboxylates and protons. In a natural ecosystem across a 2-million-year-old dune chronosequence in Jurien Bay in Australia, Hayes et al. (2014) observed that the leaf Mn concentration in non-mycorrhizal species known to release carboxylates was higher than that of co-occurring mycorrhizal species, regardless of soil age. The non-mycorrhizal carboxylate-releasing species belonged to a range of functional types: for example, species producing cluster roots, dauciform roots and sand-binding roots. In *Lupinus albus*, which also forms cluster roots, Gardner et al. (1982a,b) found that leaf Mn concentration was highly correlated with the weight of cluster roots, which are the main site for the reduction of MnO_2 in the rhizosphere.

Further to the findings in woody plants and lupin with specialised roots, Pang et al. (2018) found that the Mn concentration in mature leaves on the main stems was highly correlated with the amount of carboxylates relative to root dry weight (DW) in the rhizosheath, particularly malonate, in 100 chickpea genotypes with diverse genetic background (Fig. 3). The importance of carboxylates for Mn acquisition in chickpea significantly extends the findings in plant species with cluster roots, showing a significant contribution of carboxylate exudation to Mn uptake in chickpea plants with no cluster roots or functionally equivalent roots. Therefore, the study by Pang et al. (2018) provides strong evidence that Mn concentration in mature leaves can be used as an easily measurable proxy for carboxylates in the rhizosheath. However, the study was undertaken when chickpea was grown in a low-P environment, and whether this result would apply to plants grown at higher P levels and in soil with a high P-buffer index is still unknown and worth further research. The applicability of the relationship in other crop species needs to be verified and studies are underway in *Glycine max* and *Brassica napus* in research groups in China. Leaf Mn concentration as a proxy for belowground carboxylates offers an important breeding trait, and might provide a valuable tool for the screening of a large number of plants in breeding programmes.

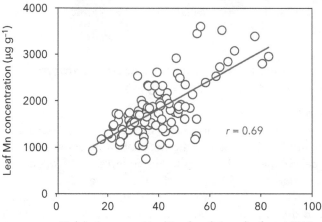

Figure 3 Correlation between manganese (Mn) concentration in mature leaves on the main stems and the amount of malonate in the rhizosheath soil per root dry weight for 100 chickpea genotypes with diverse genetic background, after growing for 7 weeks in washed river sand supplied with 10 µg P g^{-1} dry soil as FePO$_4$, which was growth-limiting. Adapted from Pang et al. (2018).

Given that the species in a mixture share the consequences of rhizosphere modification, it is difficult to separate the effect of each species; however, leaf manganese concentrations ([Mn]) can be used as a proxy for interspecific facilitation via rhizosphere carboxylate release (Lambers et al., 2015a, Pang et al., 2018, Wang and Lambers, 2019). If there is a greater leaf [Mn] of non-P-mobilising species in intercropping/mixtures than that in monoculture, it would provide evidence for facilitation via carboxylate release of P-mobilising species. For example, greater leaf [Mn] of wheat was observed when grown with white lupin, which increased with increasing density of lupin (Fig. 4; Gardner and Boundy, 1983). When grown with *Carex korshinskyi*, a forb that releases a large amount of carboxylates, neighbouring species such as *Leymus chinensis* and *Stipa grandis* usually had a greater leaf [Mn] and aboveground biomass (R.-P. Yu, unpublished data), providing compelling evidence that interspecific facilitation via carboxylate release prevails on P-impoverished soils.

3.5 Advances in phosphatases in P acquisition

Soil organic P represents 40–80% of the total P pool in soils (Menezes-Blackburn et al., 2013). Plants cannot take up organic P directly, but rather, organic P is first hydrolysed to orthophosphate by root or microbial-released phosphatases (Tarafdar and Jungk, 1987, Richardson et al., 2011). Different enzyme types are involved in mineralisation of organic P, and these have considerable differences in hydrolysis efficiency (Menezes-Blackburn et al.,

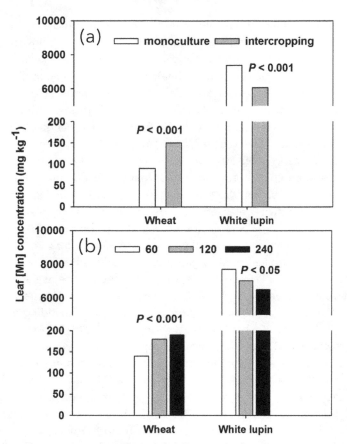

Figure 4 Leaf manganese concentrations ([Mn]) of wheat and white lupin in (a) monoculture (sowing rate: wheat 85 kg ha⁻¹, white lupin 180 kg ha⁻¹) and intercropping (sowing rate: wheat 85 kg ha⁻¹, white lupin 135 kg ha⁻¹), and (b) intercropping with different sowing rate of white lupin (60, 120 and 240 kg ha⁻¹). Bars represent mean values (n=4). P value indicates difference between (a) monoculture and intercropping, and (b) different sowing rate of white lupin (LSD). Adapted from Gardner and Boundy (1983).

2013). Some phosphatases and carboxylates have a synergistic effect on organic P mineralisation. Clarholm et al. (2015) summarizes that citrate or oxalate chelates and removes metal anions from the supramolecule aggregate or soil organic matter molecules first, and then phosphatases (e.g., acid phosphatase) hydrolyse organic P into inorganic P from newly exposed materials. Among phosphatase types, phytases are capable of initiating the cleavage of phytate molecules, making P plant-available (George et al., 2007a). Rapid utilisation of phytate-P in soil also needs the synergistic effects of carboxylates to mobilise phytate (Dissanayaka et al., 2015, Giles et al., 2017, 2018, Maruyama and Wasaki, 2017).

Soil zymography, an in situ non-destructive method combined with fluorescent substrates and ^{14}C imaging, has been used to investigate the two-dimensional distribution of phosphatases (acid and alkaline) and the interaction with microbes in the rhizosheath (Spohn and Kuzyakov, 2013, Razavi et al., 2019). The results show a spatial separation among acid and alkaline phosphatase activity and hotspots of bacteria in the rhizosheath in response to different soil P levels (Spohn et al., 2015), P sources (Ma et al., 2019), root exudate composition (Zhang et al., 2019d), growth stages (Ma et al., 2019) and plant species (Razavi et al., 2016) in rhizobox studies. This method reveals the linkages between spatial and temporal dynamics of enzymes and microbes for organic P mobilisation (Kuzyakov and Razavi, 2019). A unified visualisation of phosphatase activity released by roots or microbes, coupled with other imaging techniques and analytical methods shines light on hotspot and hot moments in the rhizosphere (Kuzyakov and Razavi, 2019, Razavi et al., 2019).

Genetically modified (GM) organisms will likely enhance the capacity of plants to use organic P. For example, greater acid phosphatase activity and P content were observed when *LASAP2* was overexpressed in *Nicotiana tabacum*, tobacco (Wasaki et al., 2009). Overexpression of phytase-encoding genes such as *OsPHY1* (Li et al., 2012) or *GmPAP4* (Kong et al., 2014) in plants enhanced plant growth and P acquisition when supplied with phytate as a P source. Enhanced biomass and P content in *Arabidopsis* were observed by transgenic expression of *MtPAP1* from *Medicago truncatula* (Xiao et al., 2006). Similarly, introduction of a phytase gene, *MtPHY,* in *M. sativa* improved plant performance when supplied with phytate (Ma et al., 2012). However, most studies on GM plants have been conducted in agar or in pots in short-term experiments, rather than field conditions (George et al., 2005, 2007b). The efficiency of GM plants to mobilise organic P depends on the P source and amount of the expressed phosphatase or released carboxylates, as well as whether the exudates are close to site of organic P mineralisation. This is because of the low mobility of phosphatase, and the required spatial and temporal vicinity needs further study (Giles et al., 2016, 2017, 2018).

3.6 Enhancing storage of P in root cells to improve fertiliser-use efficiency

For crops, P fertiliser is often applied as a band below the seed at sowing (McLaughlin et al., 2011). However, the proportion of fertiliser P taken up by the crop can be low due to losses prior to plant uptake, because of reaction with soil components to form poorly soluble inorganic compounds (especially in acid soils), uptake by microbes, and, in high-rainfall areas with sandy soils, run-off and leaching (McLaughlin et al., 2011, Simpson et al., 2015). Faster uptake by roots would reduce these losses. However, a factor limiting the uptake of

P by seedling roots is, presumably, high-tissue P concentrations. Phosphorus is stored in seeds primarily in the form of phytate (Raboy, 2009), an organic macromolecule, which allows safe storage of high concentrations of P without impacting cell osmotic pressure. After germination, the seedling quickly (within days) converts phytate-P in the seed back to inorganic P, while also commencing P uptake from the soil (Ryan et al., 2019b, Nadeem et al., 2011, 2012). Hence, young seedlings high in leaf P take up P from soil including fertiliser P, but are no longer able to store P as phytate. If fertiliser P could be absorbed faster by a seedling (and safely stored for future use), then less fertiliser P would be lost prior to root uptake, and hence fertiliser rates could be reduced. However, this requires seedlings with high-tissue P concentrations to quickly absorb and safely store large amounts of P.

The mechanism of P regulation in plant and seedling tissues (to avoid deficiency or toxicity) is storage of P in cell vacuoles: this has been assumed to be the only significant mechanism (Lee and Ratcliffe, 1993, Liu et al., 2016b). However, there is a second mechanism for P storage in roots that involves plastids (Ryan et al., 2019b), validating a suggestion by Brodelius and Vogel (1985). Ryan and co-workers showed that roots of seedlings of a pasture legume, subterranean clover, contained globular structures, similar to the globoids that contain phytate in seeds (Lott et al., 1978), and that these exhibit an extraordinarily high P concentration (up to and > 3000 mmol kg^{-1}). These do not contain phytate, but precipitated/crystallised salts of P. When roots were exposed to additional P, they stored it in these structures. Molecular analyses at whole transcriptome resolution combined with high-resolution electron microscopy indicated that the globular structures were plastids (Ryan et al., 2019b). This work led to the hypothesis that regulating the key genes controlling P movement through the root into the plastids could enhance uptake of P from P fertiliser, and thereby improve the P-uptake efficiency from P fertiliser and reduce fertiliser application rates (Fig. 5). Further work in this area is warranted.

4 Arbuscular mycorrhizal fungi (AMF) and fine root endophytes

The many ways in which the symbiosis between crops and root-colonising fungi, AMF, can impact crop growth are reviewed elsewhere in this volume. The literature pertaining to the need to manage occurrence of AMF in agricultural systems with a view to maximise crop yield was reviewed by Ryan and Graham (2018). Several of their key points are reiterated here. One was, that much of the literature lacks applicability to farming systems due to weaknesses with the design and interpretation of glasshouse experiments (see also Section 7 below) and a lack of understanding of the commercial agricultural systems to

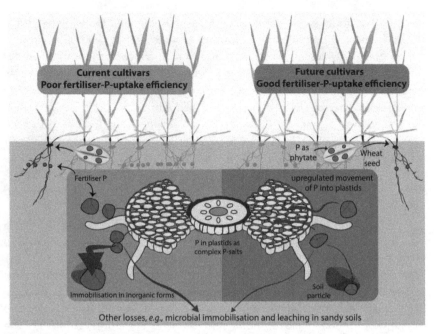

Figure 5 Schematic drawing showing how a cultivar with enhanced capacity to store phosphorus (P) in root plastids could reduce losses of fertiliser P, and hence reduce P-fertiliser requirements for wheat growing in an acid soil.

which results were extrapolated, and this extends to poor understanding of basic agronomy field techniques and science. While controversial (see Rillig et al., 2019), the failure of a very large body of literature to influence crop breeding or farm management suggests that there is a need to consider their arguments. A second key point made by Ryan and Graham (2018) and reinforced by Ryan et al. (2019a) is the need to apply a systems agronomy approach, that is, the benchmarking of credible physiological crop production limits in order to understand and remove the constraints in achieving them with the aim to develop regionally adapted, resource-efficient agronomic practices. Hence, research on AMF and crop-P nutrition requires the contribution of AMF towards P uptake and crop yield to be explored within the framework of the crop critical external P requirement. Research in this context is scarce. However, two excellent examples came from China (Deng et al., 2017, Mai et al., 2018). Deng et al. (2017) grew a wheat-maize rotation over two years with six rates of P addition. They determined the external critical soil P concentration, and found that this corresponded to a relatively high level of colonisation by AMF (Fig. 6). This work suggests that for a farming system with a target soil P sufficient to meet the crop external critical P requirement, there may be no need to alter management to enhance colonisation by AMF. This is a useful finding as

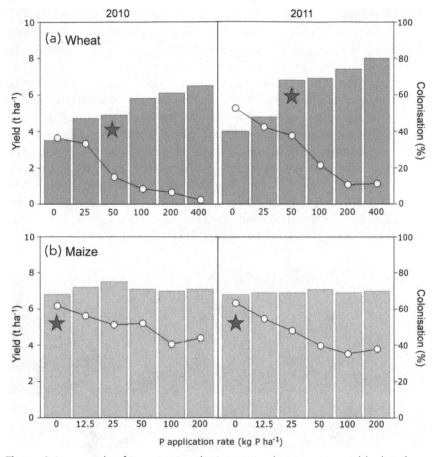

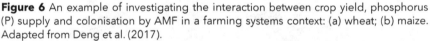

Figure 6 An example of investigating the interaction between crop yield, phosphorus (P) supply and colonisation by AMF in a farming systems context: (a) wheat; (b) maize. Adapted from Deng et al. (2017).

commercial inoculants are not widely used (Ryan and Graham, 2018, Hart et al., 2018).

Suggestions that crop breeders should consider how to enhance the level of colonisation and the mycorrhizal contribution to P uptake (e.g. Martin-Robles et al., 2018) need to be critically appraised in the context of the need to have a target soil P that meets crop external critical P requirements. In addition, it should be considered that higher colonisation may have unintended negative consequences in some circumstances such as reduced yield (Ryan et al., 2005) or promotion of plant parasitic nematodes (Frew et al., 2018). Moreover, attempts to change other root traits to enhance P uptake may require consideration of AMF. For instance, while, functionally, AMF were found by Jakobsen et al. (2005) to substitute for root hairs on mutant barley plants lacking root hairs, they were suggested to possibly override the P-acquisition benefits of selecting for longer root hairs in white clover (Caradus, 2012). The trade-offs that can occur at the genotype or individual plant level due to the carbon costs of colonisation by AMF and other roots traits are discussed in Section 5.

An interesting finding, whose significance for root uptake of P is yet to be explored, is that by Orchard et al. (2017a) that the 'fine root endophyte' previously thought a minor species of AMF, are in fact members of a different subphylum than the AMF, the Mucoromycotina. Orchard et al. (2017b) suggested that they belong to the Densosporaceae family within the Endogonales. Hence, it seems that plants contain two groups of arbuscule-forming endophytes that may be involved in plant P nutrition. Fine root endophytes (FRE) are relatively little researched, but are prolific in some agricultural systems including pastures in New Zealand and dryland cereal crops and pastures subjected to seasonal waterlogging in Australia (Orchard et al., 2017b). They may also enhance host

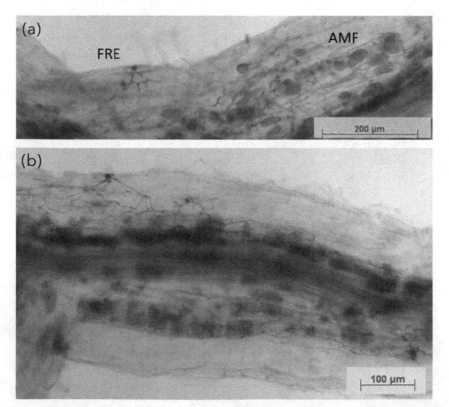

Figure 7 The arbuscule-forming root symbiont, fine root endophyte (FRE), colonises roots either in close association with arbuscular mycorrhizal fungi (AMF) (a) or by itself (b); its role in plant root phosphorus (P) uptake remains to be explored. Photograph by Jeremy Bougoure.

plant P uptake (e.g. Crush, 1973) and respond negatively to increasing P supply in a way similar to AMF (Jeffery et al., 2018). It is likely that the literature on AMF in agricultural systems has been confused by researchers either assuming FRE were AMF when scoring colonisation using light microscopy or missed their presence when using molecular techniques to characterise populations. FRE and AMF often colonise the same plant and even co-habitat the same sections of roots (Fig. 7), untangling their relative contribution to plant root P uptake will likely prove challenging. Beyond FRE, there is also an increasingly complex range of interactions between AMF and other soil biota (Ryan and Graham, 2018) including those closely associated with AMF, both internally (Salvioli et al., 2016) and externally (Zhang et al., 2014b, 2019b). However, harnessing such interactions for agricultural purposes will be a long-term endeavour (Ryan and Graham, 2018).

5 Trade-offs in traits

Roots exhibit diverse functional traits, which allow them to acclimate to low-P soils and to increase soil P acquisition including changes in root mass allocation, root morphology, P-mining exudates and mycorrhizal symbioses (Vance et al., 2003, Lambers et al., 2006, Shen et al., 2011). All these strategies pertain to different soil P sources in contrasting ways, presumably with different costs (carbon and other resources) and benefits (Lynch and Ho, 2005, Ryan et al., 2012, Raven et al., 2018). Thus, a plausible hypothesis is that the cost of each adjustment may limit the ability of plants to express all strategies simultaneously (Ryan et al., 2012, Nazeri et al., 2014). Indeed, plant species seem to show contrasting ways to express these root functional traits (Lambers et al., 2008, Ryan et al., 2016, Lyu et al., 2016). For instance, crops such as maize and wheat exhibit stronger root morphological responses than physiological responses to P deficiency (Lyu et al., 2016, Deng et al., 2014, Wen et al., 2017). Other crops such as chickpea and faba bean show a limited root morphological response to varying P supply, but a significant modification in root exudation (Liu et al., 2016a, Lyu et al., 2016, Zhang et al., 2016). Plant species that are adapted to either relatively fertile soils (e.g. Brassicaceae) or severely P-impoverished soils (e.g. Proteaceae), tend to be non-mycorrhizal (Lambers et al., 2008, 2015b, Brundrett, 2009), but species that are adapted to less extreme environments exhibit all three groups of root functional traits (i.e. root morphology, P-mobilising exudates and mycorrhizal symbioses) with an intermediate strategy. An interesting question is how plants coordinate the different root functional traits to enhance soil P acquisition under contrasting soil P availability.

Great progress has been made in revealing the evidence associated with coordination and trade-offs among root functional traits within or among species, linked with their nutrient-acquisition strategies, environmental gradients and evolutionary history (Reinhart et al., 2012, Waddell et al., 2015, Ryan et al., 2012, Roumet et al., 2016, Li et al., 2017, Ma et al., 2018, Wen et al., 2019). Generally, thin roots with a relatively high-specific root length, are associated with rapid growth rates and low root tissue density, but short life span, indicating a resource-acquisitive strategy. In contrast, thick roots with a relatively low specific root length are linked with slow growth rates and high root tissue density, but long life span, suggesting a resource-conservative strategy (Valenzuela-Estrada et al., 2008, McCormack et al., 2015, Freschet and Roumet, 2017). At the species level, this is particularly prevalent in natural forest and grassland ecosystems, and species with thinner roots (thin-root species) always have lower colonisation by AMF, but higher root branching, whereas species with thicker roots (thick-root species) generally exhibit higher colonisation by AMF, but consistently lower root branching (Liu et al., 2015, McCormack et al., 2015, Chen et al., 2016, Li et al., 2017). Wen et al. (2019) considered root exudation together with root morphology and mycorrhizal symbioses in shaping P-acquisition strategies of 16 crops, with a range of trade-offs among these root-functional traits. Taken together, in response to a low P availability, species with thinner roots mainly depended on their root morphology (e.g. high root-branching intensity, high specific root length) to enhance nutrient (including P) acquisition by increasing soil exploration. Conversely, species with thicker roots relied more on higher colonisation by AMF to compensate for a low root-absorptive surface and/or more P-mobilising exudates (e.g. significant carboxylate exudation) to mine sparingly soluble P in the rhizosheath (Kidd et al., 2015, Wen et al., 2019).

According to the principle of carbon economy, optimisation of internal resource allocation should occur, underpinning the adjustments of various root functional traits at the individual plant level, towards a balance between costs and benefits (Lynch and Ho, 2005). Supporting this idea, Ryan et al. (2012) found for 10 *Kennedia* species that under low-P treatment, inoculation with AMF decreased the amount of carboxylates in their rhizosheath that likely resulted in inoculated plants accessing a greater proportion of soil P from a labile inorganic P pool, rather than sparingly available P sources. Compared with plants without inoculation, association with AMF can also modify root architecture and morphology (Hetrick et al., 1991, Atkinson et al., 1994, Berta et al., 1995, Yang et al., 2015a). However, the association between root architecture and mycorrhizal growth response remains contentious (Brundrett, 2002, Maherali, 2014, Smith and Read, 2008). Generally, colonisation by AMF decreases the root mass fraction which may contribute to more carbon diversion to mycorrhizal symbionts, and thereby less carbon being allocated to root growth (Berta et al., 1995, Johnson et al., 1997, Graham and Abbott, 2000, Ryan et al., 2016). It also decreases specific root length (Berta et al., 1995, Mendoza, 2001, Sun and Tang, 2013), but increases root tissue density

(Zangaro et al., 2007), which suggests the root morphology of inoculated plants becomes more conserved. Not all the effects are due to improved P nutrition, because similar results can be found in high-P conditions (Berta et al., 1995, Zangaro et al., 2007, Maherali, 2014).

Overall, the findings show that, at the species level and at individual plant level, root morphology, P-mobilising exudates and mycorrhizal symbioses together shape P-acquisition strategies, with a range of trade-offs among these root functional traits.

6 Microbially-mediated efficient P acquisition in species-diverse systems

Soil micro-organisms play a pivotal role as P sinks and sources, and mediate key soil P processes such as P mineralisation and immobilisation (Oberson and Joner, 2005). The density and diversity of soil micro-organisms are particularly great in the rhizosphere. Plants can affect rhizosphere micro-organisms to mobilise soil P resources, and thus enhance P acquisition through exuding C-rich and diverse rhizodeposits (Badri et al., 2009, Drogue et al., 2012). Yet, it remains unclear whether and how soil/rhizosphere micro-organisms can mediate plant P acquisition in species-diverse systems such as natural grasslands and intercropped agroecosystems.

Soil microbial P pools represent an important source of P for plants (Hinsinger et al., 2015, Bünemann et al., 2013), because most of the microbial P is located in labile intracellular compounds, and the turnover of microbial P is often fast, for example, in grassland soils (Oberson and Joner, 2005). In natural grasslands, plant diversity increases soil microbial activity (Chung et al., 2007, Eisenhauer et al., 2010, Zak et al., 2003). This greatly contributes to the increase in overall plant production (Zak et al., 2003, Hooper et al., 2005), which generally increases with increasing number of plant species grown together. This suggests that systems where multiple species are grown together may enhance soil microbial P pools and turnover, and thus facilitate plant P acquisition. In agroecosystems, intercropping alters soil microbial biomass and composition compared with monocropping systems (Song et al., 2007, Tang et al., 2014, Wang et al., 2007). For example, Tang et al. (2014) showed that both durum wheat/chickpea and durum wheat/lentil intercropping promoted P storage in soil microbial biomass compared with the respective monocrops. In addition to microbial biomass, cereal-legume intercropping (e.g. wheat/soybean) increased root fungal diversity of intercropped wheat which was correlated with shoot P content (Bargaz et al., 2017). However, increased soil microbial P pools in species-diverse systems do not always translate into enhanced plant P acquisition and productivity (Li et al., 2010), suggesting that other mechanisms may also determine the soil P turnover and subsequent P acquisition.

Phosphate-solubilising micro-organisms including bacteria, fungi and actinomycetes promote plant growth by transforming sparingly soluble P into soluble P forms in different production systems (Khan et al., 2014). Intercropping rubber (*Hevea brasiliensis*) with banana (*Musa acuminata*) or cassava (*Manihot esculenta*) significantly enhanced the diversity of phosphate-solubilising bacteria, which was associated with increased P solubilisation (Sungthongwises, 2016). Enhanced activity of phosphate-solubilising bacteria can be attributed to the exudation of lectins by crops which positively influence plant growth-promoting rhizobacteria (Schelud'ko et al., 2009). In addition, phosphate-solubilising bacteria such as Gram-negative *Pseudomonas fluorescens* may exude antibiotics (Yang et al., 2008, Taurian et al., 2010), providing protection for plants against soil-borne pathogens, which could influence plant health and subsequent plant P acquisition (Singh et al., 2010).

Future work should not only investigate microbial biomass or coarse-scale microbial composition in species-diverse systems, but also identify specific microbial functional groups, and their role and interactions with plant species. In addition, bacteria-feeding nematodes may increase soil phosphatase activity and soil P availability through grazing bacteria (Djigal et al., 2004). It might be worthwhile to examine how intercropping affects soil fauna such as earthworms and nematodes, and their role in soil P cycling.

7 Farming-management practice in P acquisition at the intensive agricultural system level

There are many reviews and summaries on soil- and crop P-management strategies (Shen et al., 2013, Withers et al., 2014, Zhang et al., 2019c). The beneficial management practices include developing new fertilisers, shaping root-microbe interactions and rhizosphere management (Table 1). Here, we emphasise the role of root functional traits in improving P acquisition in the context of intensive agricultural systems, especially under high P input and high plant-density conditions.

The importance of root systems in supporting plant growth and P uptake has been extensively studied (Teng et al., 2013, Wen et al., 2017, Jeffery et al., 2017b, Haling et al., 2018, Pang et al., 2018). However, many of these studies were performed under P-deficient conditions and/or on individual plants (Pearse et al., 2007, Lyu et al., 2016, Wen et al., 2017, Pang et al., 2018). Few have considered the context of intensive agricultural systems, which are characterised by high resource input (including P fertiliser) and high plant density (Shen et al., 2013, Withers et al., 2014, Mi et al., 2016). Under such a condition, with a large quantity of fertiliser applied, crop growth is typically not limited by nutrients (Withers et al., 2014). Moreover, high plant density potentially leads to high root-root competition, so that many root traits that

Table 1 Examples of phosphorus (P)-management practices that have the potential for development and application in cropping systems

Management practices	Description	Comments	References
Novel/smart P fertilisers	Considering nutrient-release rate matches crop P demand, soil properties and climatic feature	Integrate knowledge in plant science, agrochemicals and materials; the high cost often limits the application in grain crops	Calabi-Floody et al. (2018); Weeks and Hettiarachchi (2019)
Microbial fertiliser (bio-fertiliser, bioformulation fertiliser)	Shaping beneficial root-microbial relationships to promote plant growth and P uptake	The functions involved in production and regulation of phytohormones, release of nutrients (P) to plants and control of pathogens; the cost may be high and the beneficial effects may be unstable under changing soil conditions	Richardson et al. (2011); Nadeem et al. (2014); Calabi-Floody et al. (2018)
Intercropping (mixed cropping)	Two or more crop species or genotypes with niche complementarity, growing together and coexisting for a prolonged time	Improve plant diversity; suitable for small-hold farming and low-input/resource-limited agricultural systems, but may not be suitable for intensive cropping systems and mechanisation	Li et al. (2014b); Brooker et al. (2015)
'4R' nutrient stewardship	Applying the right source, at the right rate, right time and in the right place	It supports multiple ecosystem services; the challenge for 4R stewardship is to develop an effective management package that works within a site-specific, dynamic, and complex system	Bruulsema et al. (2019); Grant and Flaten (2019)
Soil-based P-management strategies	Based on soil P test and yield response; considering a nutrient balance of P fertiliser input and plant uptake	Applying fertilisers to build-up and maintain an 'insurance' soil P level for maximum crop growth and yield	Li et al. (2011); Withers et al. (2014)
Plant/rhizosphere-based P-management strategies	Maximising root/rhizosphere efficiency to enhance soil P acquisition	Enhance the utilisation of legacy soil P and increase fertiliser-P recovery by rhizosphere management	Li et al. (2011); Shen et al. (2013); Zhang et al. (2019c)

are advantageous for P uptake in individual plants, may not be suitable for a population in intensive cropping systems (Weiner et al., 2010, York et al., 2015, Mi et al., 2016, Zhu et al., 2019). So, for sustainable cropping systems and P management, we should reconsider the roles of root functional traits in enhancing P acquisition to combine the root functional traits of high P input (high legacy soil P reserves) and high plant density.

The improvement of crop productivity has relied heavily on the input of chemical fertilisers including P (Li et al., 2011, Shen et al., 2011, Withers et al., 2014, Zhang et al., 2019c). The continuous overuse of P not only resulted in a significant increase in plant-available P concentrations, but also led to the accumulation of a largely unavailable soil P pool (soil P reserves) (MacDonald et al., 2011, Sattari et al., 2012). How to make more efficient use of both applied P and the soil P reserves, and reduce target soil P levels has become increasingly urgent for the development of sustainable cropping systems. Soil P availability significantly affects the response intensity and the effectiveness of root functional traits. Generally, adjustments of root morphology, release of P-mobilising exudates and colonisation by AMF are enhanced by a low shoot P status and low soil P availability, and suppressed by increasing plant P status and soil P availability (Deng et al., 2014, Wang et al., 2016, Wen et al., 2017). Supporting this view, Teng et al. (2013) found that the optimal P supply for maximum grain yield of wheat coincided with the critical P concentration for P-starvation marker gene responses in roots, and was at or near the change points for downregulating root responses to P supply. At an optimal soil P level, grain yield was highest, but also maintained a relatively high response intensity of root traits and a relatively high expression level of genes encoding P transporters and phosphatases (Teng et al., 2013). Similar results were found in maize in response to increasing P supply in the field (Deng et al., 2014). Overall, in the context of intensive agricultural systems, it is an important strategy to reduce or withhold P applications, while using the soil P reserves for crop growth, until soil P concentrations decline to the optimum level for maximum crop productivity (critical soil P level). That means, improving P-use efficiency and crop production can be achieved through enhancing root/rhizosphere efficiency with lower external P input.

Increasing plant density is one of the most important ways to increase yield per hectare (Duvick, 2005, York et al., 2015). When plant density is too low, each individual plant may perform at its maximum capacity, but there are not enough plants to reach the maximum yield per hectare (Shao et al., 2018). In contrast, if the plant density is too high, plants may compete with each other, and the performance of individual plants becomes a limiting factor for maximum crop yield (Xu et al., 2017, Shao et al., 2018). Thus, it is important to use appropriate plant densities to reach a high yield. Understanding root responses, and thereby P uptake, to high planting density will provide important information for better crop management in the context of intensive agricultural systems. In response to increasing plant density, plants exhibit significant modifications in root growth and spatial distribution such as a marked reduction in overall root size, root mass and total root length. These changes have an indirect effect on plant nutrient and water uptake (Chen et al., 2013, York et al., 2015, Shao et al., 2018). Different genotypes show contrasting tolerance to high plant density, opening up the possibility of selecting elite genotypes with high yield performance under high plant density (York et al., 2015, Mi et al., 2016). Comparing eight maize hybrids under three plant densities, Shao et al. (2019) found that genotypes with less variation in root size, medium root size, medium broad-root system and more roots distributed in the inter-row space reduced root competition between neighbouring plants. These hybrids tended to achieve higher yield at high plant density. In summary, plants may show shifts in P-acquisition strategies at the individual plant level and at the community level. Aiming to optimise yield performance and nutrient-acquisition efficiency at the community level, we should consider root functional traits to improve P acquisition in the context of high plant density.

Screening genotypes for root traits may be done best under high-density conditions that replicate those of the target crop. Glasshouse experiments in spaced pots where the canopy can expand beyond the edge of the pot or spaced single plants or spaced rows in the field may not properly replicate these conditions. For instance, when Jeffery et al. (2017a) grew annual pasture legumes in a glasshouse with and without reflective sleeves to mimic sward conditions, the addition of the sleeves reduced shoot dry mass by 10–28%. However, their impact on roots was much greater, being a 39–59% decrease in root dry mass, and an increase in specific root length. Most notably, there was a 3.5–12-fold increase in the amount of rhizosheath carboxylates expressed relative to unit root length.

8 Breeding to improve P acquisition

A review by Wang et al. (2019) summarised the parental lines, types and sizes of mapping populations used to identify quantitative trait loci (QTLs) associated with P efficiency in various crops including rice, maize, wheat, common bean, soybean and oilseed rape. With the rapid development of high-throughput sequencing technology, and the construction of high-density genetic maps, genome-wide association studies (GWAS) has been increasingly used for the mapping of P-efficient genes, for example, in soybean (Ning et al., 2016) and oilseed rape (Wang et al., 2017a). Phosphorus-efficiency traits used for QTL mapping studies primarily involve four groups: (1) root traits improving soil inorganic P availability such as acidification of the rhizosphere and exudation of carboxylates, hydrolytic enzymes, for example, acid phosphatases, ribonuclease, phytase; (2) traits improving P uptake such as root architecture and morphology, AMF and beneficial rhizosphere micro-organisms; (3) traits improving the physiological P-utilisation efficiency; and (4) traits associated with yield production (Wang et al., 2019).

Though numerous QTLs affecting P efficiency have been identified in key crops, few causal genes have been identified, and the progress in the breeding of P-efficient crops using marker-assisted selection (MAS) has been limited (Wang et al., 2019). However, there is a very successful example in rice. One major QTL cluster, *Pup1* (phosphorus uptake1) has been identified and introgressed into several rice varieties using a marker-assisted backcrossing approach, leading to a dramatic increase in P-acquisition efficiency and enhanced yield of those lines when grown under a suboptimal P supply (Chin et al., 2011). Functional characterisation of a *Pup1* QTL in rice identified a protein kinase gene (*PSTOL1*, phosphorus starvation tolerance 1), which increased adventitious root development under low P availability and conferred low-P tolerance (Gamuyao et al., 2012). Cloning of *PSTOL1* has been a major breakthrough in a P-efficient rice breeding programme. Further to the findings in rice, several *PSTOL1* genes were identified in sorghum (*Sorghum bicolor*) which were associated with P efficiency, attributed to larger root systems with longer and finer roots and more laterals (Hufnagel et al., 2014). Leiser et al. (2014) found that *PSTOL1* homologs consistently enhanced sorghum P efficiency between two different association panels in Africa and Brazil with different genetic backgrounds and environments, suggesting the potential for using markers linked to *PSTOL* genes for molecular breeding to improve P efficiency in sorghum. Another interesting example, in soybean, is the identification of a QTL, both in GWAS and linkage mapping for a candidate gene-encoding acid phosphatase, *GmACP1* (Zhang et al., 2014a). As P acquisition is a complex trait due to the significant plant-soil interaction, the lack of understanding of P uptake is a bottleneck for both conventional and GM approaches. The availability of complete genome sequences for more crops, and the combination of conventional linkage mapping, association mapping, QTL-sequencing, transcriptomics and gene-editing technologies will speed up the identification of genes underlying QTLs associated with P efficiency. Further progress is required using gene-linked markers to improve nutrient-acquisition efficiency and crop yield. The identification of these genes would provide the opportunity to improve P efficiency through MAS or gene manipulation.

For the successful breeding of P-efficient crops, high-throughput selection of genotypes is required. The development of high-throughput plant phenomics facilities has enabled accurate and standardised investigation of P-efficiency-related traits (Furbank and Tester, 2011, Ubbens and Stavness, 2017). So far, much of the reported work comes from glasshouse or hydroponic studies. Although field assessments would be the ultimate test of the superiority of new varieties, controlled environments reduce environmental variation, thus revealing traits associated with P efficiency. Assessment of P efficiency in the field is often problematic due to the large spatial heterogeneity of soil P, masking genetic effects leading to rather low heritability (Van De Wiel et al., 2016). As plant growth and P uptake rely on a complex interaction of plant, soil and rhizosphere (Bovill et al., 2013), traits for P improvement identified under controlled conditions often fail to show advantages in the field, with significant variation among phenotypes in different culture systems. For example, using three culture systems (i.e. hydroponics, vermiculite and paper rolls) for the phenotyping of eight root traits in a maize recombinant inbred-line population at seedling stage, Liu et al. (2017) found that the root traits in hydroponics and vermiculite were weakly correlated with P-uptake efficiency in the field, while there was no correlation for plants grown in paper rolls. High-throughput screening is desirable, but if the results do not correlate with those obtained in the field their application will be limited.

In many crops, selection for P efficiency has been slow, partly due to inconsistent definitions of P efficiency. Often, different terms are used, even though the efficiency is calculated in the same way, leading to different conclusions about genetic differences in nutrient efficiency depending on the criterion of efficiency used (Bovill et al., 2013). More consistent use of terminology and definitions for improving P efficiency are needed. Both single traits and relative traits (the ratio of the value of a trait for plants grown at a reduced P supply to that of plants grown with optimal P supply) have been used for the identification of QTLs for the effects of P supply on yield, rhizosphere acidification and biomass-related traits. While single traits are often commonly used for QTL associated with P-efficiency traits, relative traits indicate a genotype's tolerance to lower P supply. For example, in barley Gong et al. (2016) reported that a QTL for P responsiveness was mapped to the same locus as a QTL for yield at 30 kg P ha^{-1}, suggesting that P responsiveness and grain yield could be improved simultaneously under high-input cropping systems. Therefore, it would be more useful to identify the co-located QTLs, both for a single trait and for a relative trait in breeding programmes (Wang et al., 2019).

9 Case study: mobilisation of phosphorus and manganese in cluster roots of *Lupinus albus*

While working towards his PhD, Bill Gardner was the first to show that cluster roots of *Lupinus albus*, a non-mycorrhizal species, release citrate, providing a mechanism for their highly-efficient P mobilisation (Gardner et al., 1981). The authors also showed that these cluster roots were equally efficient at mobilising Mn, and followed up their laboratory studies with field experiments, demonstrating that intercropped wheat benefitted from the P-mobilisation capacity of *L. albus* and showed higher leaf [Mn] (Gardner and Boundy, 1983).

Much later, leaf [Mn] was proposed as a proxy for carboxylate concentrations in the rhizosheath (Lambers et al., 2015a), and as a tool to screen for carboxylate release in *Cicer arietinum* (Pang et al., 2018). Mycorrhizal strategies are effective at low soil P availability, but when the P availability is very low, for example, in strongly P-sorbing soils, a strategy based on carboxylate release is more effective (Lambers et al., 2018, Raven et al., 2018, Parfitt, 1979). That would account for lupins being a successful crop on both sandy soils in Australia (Gladstones, 1970, Lambers et al., 2013) and strongly P-sorbing volcanic soils in Chile (Lambers et al., 2013, Baer, 2006).

As far as we know, all *Lupinus* species are non-mycorrhizal, but only a small number produce cluster roots, and the ones that do not, likely all release carboxylates (Lambers et al., 2013). Bolland (1997) compared four *Lupinus* species in a field experiment, showing that the two with cluster roots (*L. albus* and *L. cosentinii*) grow better and produce more grain at a low P supply than the two without cluster roots (*L. angustifolius, L. luteus*). Bolland and Brennan (2008) compared *L. angustifolius* with two crop species that do not release large amounts of carboxylates in a pot experiment using the top 0.10 m of a sandy soil that was collected under remnant vegetation that had never been fertilised and used for agriculture. To produce 90% of the maximum grain yield, *L. angustifolius* required ~67% less P than wheat, and canola required ~75% more P than *L. angustifolius*.

Cluster roots in *L. albus* are suppressed at high P supply (Gardner et al., 1983a, Keerthisinghe et al., 1998), as a result of a systemic signal associated with high leaf P concentrations (Marschner et al., 1987, Shane et al., 2003). When plants relied on symbiotic N_2 fixation, which increased their demand for P, their investment in cluster roots was greater than when they were being fed with nitrate (Wang et al., 2018b).

There is overwhelming evidence from field and laboratory studies that cluster roots are a very effective strategy to acquire P from soils with a very low P availability, both in natural and managed systems (Dinkelaker et al., 1989, Shane and Lambers, 2005). Species that released carboxylates without specialised structures were not quite as effective as those with cluster roots, but more effective than mycorrhizal species that did not release significant amounts of carboxylates (Pearse et al., 2006, Veneklaas et al., 2003). We therefore strongly disagree with a statement by Lynch (2019) that variation for the production of carboxylates is associated with P mobilisation in vitro only, and cited numerous references above on crops and native species showing that rigorous analyses have convincingly shown a benefit of such variation for P acquisition in a range of soils in the field.

Pot experiments have shown a significant P benefit of white lupin on the subsequent wheat crop (Nuruzzaman et al., 2005a,b). There is insufficient information to conclude that this effect also happens under field conditions, partly because the subsequent wheat crop was grown with an adequate P supply (Angus et al., 2015). However, in a field experiment on two acid soils of southern Cameroon, significant P benefits were shown for maize grown after some soybean cultivars, but not after others (Jemo et al., 2006). Positive effects would be expected only on soils with a low P availability, especially when these soils contain high levels of total P. Future field work should target such soils, rather than soils that have been fertilised for a very long time.

10 Conclusion

There are two contrasting soil P levels that need our urgent attention: low levels in low-input agricultural system such as many parts of Africa, and accumulation of P due to use of P fertiliser in excess of crop requirement in many developed and rapidly developing countries. It is necessary to improve the acquisition of soluble P under both scenarios, while also increasing a crop's capacity to access large pools of insoluble P. To improve P acquisition, further understanding of root traits is required including (1) improving P availability through exudation of carboxylates, protons and phosphatases; and (2) improving P uptake through changes in root architecture and morphology, AMF and beneficial rhizosphere micro-organisms. Improved crop P-acquisition efficiency would lower critical P requirements, and, therefore, reduce P-fertiliser input, while minimising soil P loss and environmental pollution.

GWAS is increasingly used for mapping P-efficient genes in a number of crops, but few causal genes have been identified. More progress is required in using gene-linked markers to improve P-acquisition efficiency and crop yield. Along with the availability of complete genome sequences for more crops, the combination of conventional linkage mapping, association mapping, QTL-sequencing, transcriptomics and gene-editing technologies would need to be combined to speed up the identification of genes underlying QTLs associated with P efficiency. Identification of these genes would provide the opportunity to improve P efficiency through MAS or gene manipulation. Although yield and P acquisition under field conditions is the ultimate aim, due to the large environmental effects on plant performance and P acquisition, trait-based selection using high-throughput phenotyping platforms with accurate and standardised protocols will be useful for the identification of P-acquisition-related traits. In addition, given the challenges of belowground sampling and observations, the development of easily measurable aboveground proxies for belowground traits will be highly desirable and ideal for the high-throughput phenotyping of large numbers of genotypes. Accompanying the genetic improvement in P acquisition, improvement in the design of cropping systems and rhizosphere management will also be required such as intercropping, rotations and modification of rhizosphere processes.

11 Future trends in research

Figure 8 shows challenges in improving P acquisition and future efforts towards this. We suggest the following trends:

i Many studies on P efficiency have been undertaken on severely P-deficient plants or soils. In many developing countries, P deficiency is a limiting factor for crop production. However, in many cereal-based systems in North America, Europe, Australia and China, there is a long history of P-fertiliser applications so that severe P deficiency is rare. However, improving P efficiency in those situations is still needed, as P recovery and P-fertiliser efficiency may be low. So far, most attempts to

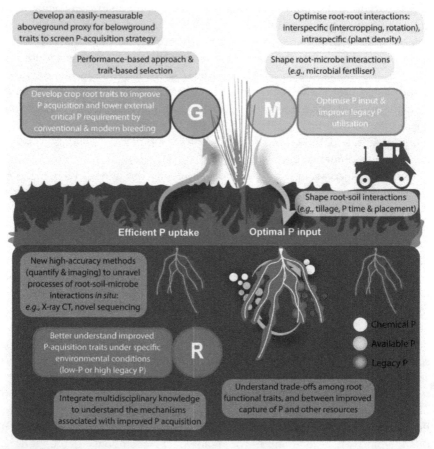

Figure 8 A summary diagram showing the challenges in improving phosphorus (P) acquisition and future efforts towards this. G, development of GENOTYPES with improved P acquisition; M, optimisation of P MANAGEMENT to increase P acquisition; R, better RESEARCH on belowground traits associated with P acquisition.

improve P efficiency have been made in low-P environment, but efforts to improve P efficiency within the context of farming systems with higher P levels must also be considered. On farms with reasonable soil P levels, selecting for P uptake under severely P-limited conditions is unrealistic; instead, systematic evaluation of putative traits for P-acquisition efficiency at different P levels is required. This would make clear whether characteristics contributing to enhanced P uptake and growth under severe P stress would still be useful at higher soil P levels. In the context of the need to maximise yields, there should be a focus on the traits that confer a decrease in the critical external P requirement of crops, and how the target critical P should be determined in a diverse system (rotated or intercropping crops) and how fertiliser practices can be best modified to ensure yield and P-use efficiency in such a system (Deng et al., 2017). Such an approach requires a good understanding of the target commercial agricultural systems, and failure to acquire and apply such knowledge can result in irrelevant research as discussed for AMF by Ryan and Graham (2018).

ii In addition to a traditional performance-based approach for yield and/ or nutrient uptake, the trait-based selection to improve nutrient uptake and yield would be useful to overcome the problem of having low heritability in the field due to often large environmental effects. With the rapid advances in molecular technology, reliable phenotyping will be key to the identification of QTLs and genes associated with P-acquisition efficiency. The development of automatic high-throughput phenotyping platforms with accurate and standardised investigation of P-acquisition-related traits will be required, as the final aim will be to improve yield and P acquisition in field. It is imperative that glasshouse screening is conducted under conditions as relevant to the field as logistically possible, with an important preliminary step being determination of the field conditions that should be replicated. For instance, presence of AMF or other soil organisms, stratification of P, environmental conditions and fluctuations, plant age or phenological stage(s) at assessment and plant density/sward conditions (Ryan and Graham, 2018, Jeffery et al., 2017a, Haling et al., 2016, 2018). Ultimately, field assessment under commercially realistic management conditions will be required.

iii Due to the difficulty of studying roots in soil under field conditions, the development of easily measurable aboveground proxies for belowground traits is highly desirable and ideal for high-throughput phenotyping of large numbers of genotypes. For example, the use of leaf [Mn] as a proxy for rhizosheath carboxylates in chickpea (Pang et al., 2018), and blumenols as shoot markers of root symbiosis with AMF (Wang et al., 2018a). More work like this on different crops and in different environments is needed to make the high-throughput phenotyping of belowground traits more accessible. In addition, new methods for root studies would be required for future root study. For example, soil zymography (Spohn and Kuzyakov, 2013) is an in-situ technique for localisation and quantification of enzyme activity in soil using thin gels with embedded substrates. It is non-destructive for soil structure and displays enzyme activities more realistically than traditional enzyme assays, which involve the destruction of soil structure.

iv There is a need to adopt better agronomic-management practices such as the design of smart farming system in intercropping and rotation to maximise P acquisition of crops. The concept of 'building up and maintenance' by Zhang et al. (2019c) suggests limiting P input in fields with high existing P levels by maintaining plant-available P in an optimal range for the yield of the target plants. This will improve the P-use efficiency of applied fertiliser, while also reducing the sorption of P fertiliser in soil. Limiting P input will also be a very effective strategy to reduce soil P loss causing environmental pollution. By optimising cropping systems and rhizosphere management such as intercropping, rotations and modification of rhizosphere processes, soil legacy P could be used more efficiently allowing a decrease of plant-available soil P to a critical level, without decreasing crop yields.

v Because P acquisition involves a number of complex traits, there is a need to develop collaboration among the disciplines of agronomy, physiology, modelling, genetics, genomics, metabolomics, proteomics, transcriptomics and bioinformatics.

12 Where to look for further information

- The Global Phosphorus Research initiative (GPRI), a collaboration between six independent research institutes in Europe, Australia and North America. http://phosphorusfutures.net.
- The Sustainable Phosphorus Alliance hosts an Annual Phosphorus Forum on topics related to phosphorus sustainability. The organisation also hosts technical webinars and workshops on phosphorus management issues. https://phosphorusalliance.org.
- International Society of Root Research (ISRR) aims to promote cooperation and communication between root researchers around the world. The society holds an international symposium every three years. https://www.rootresearch.org/.
- Rhizosphere Conferences have been held every four years around the world, providing a multidisciplinary forum for researchers in the area of the rhizosphere, to understand its complexity, and role in both natural

and agricultural ecosystem processes. https://www.rhizo5.org/home/index.html.

12.1 Review articles

- Baker, A., Ceaser, S. A., Palmer, A. J., Paterson, J. B., Qi, W., Muench, S. P. and Baldwin, S. A. (2015). Replace, reuse, recycle: improving the sustainable use of phosphorus by plants. *Journal of Experimental Botany* 66: 3523-3540.
- Cordell, D. and White, S. (2014). Life's bottleneck: sustaining the world's phosphorus for a food secure future. *Annual Review of Environment and Resources* 39: 161-188.
- Menezes-Blackburn, D., Giles, C., Darch, T., George, T. S., Blackwell, M., Stutter, M., Shand, C., Lumsdon, D., Cooper, P., Wendler, R., et al. (2018). Opportunities for mobilizing recalcitrant phosphorus from agricultural soils: a review. *Plant and Soil* 427: 5-16.

13 Acknowledgement

Zhihui Wen was supported by the China Scholarship Council (CSC). Daniel Kidd was funded by Meat and Livestock Australia (MLA) and Australian Wool Industry (AWI). We acknowledge OOid scientific for the preparation of Figs 5 and 8.

14 References

Angus, J. F., Kirkegaard, J. A., Hunt, J. R., Ryan, M. H., Ohlander, L. and Peoples, M. B. 2015. Break crops and rotations for wheat. *Crop and Pasture Science* 66(6), 523-552.

Atkinson, D., Berta, G. and Hooker, J. E. 1994. Impact of mycorrhizal colonisation on root architecture, root longevity and the formation of growth regulators. In: Gianinazzi, S. and Schüepp, H. (Eds) *Impact of Arbuscular Mycorrhizas on Sustainable Agriculture and Natural Ecosystems*. Basel: Birkhäuser Basel. pp. 89-99.

Badri, D. V., Weir, T. L., Van Der Lelie, D. and Vivanco, J. M. 2009. Rhizosphere chemical dialogues: plant-microbe interactions. *Current Opinion in Biotechnology* 20(6), 642-650.

Baer, E. V. 2006. Relevant points for the production and use of sweet lupin in Chile. México, where old and new world lupins meet. Proceedings of the 11th International Lupin Conference, Guadalajara, Jalisco, Mexico, 4-9 May 2005. International Lupin Association. pp. 116-119.

Barber, D. A. and Lynch, J. M. 1977. Microbial growth in the rhizosphere. *Soil Biology and Biochemistry* 9(5), 305-308.

Bargaz, A., Noyce, G. L., Fulthorpe, R., Carlsson, G., Furze, J. R., Jensen, E. S., Dhiba, D. and Isaac, M. E. 2017. Species interactions enhance root allocation, microbial diversity and P acquisition in intercropped wheat and soybean under P deficiency. *Applied Soil Ecology* 120, 179-188.

Bayuelo-Jiménez, J. S., Gallardo-Valdéz, M., Pérez-Decelis, V. A., Magdaleno-Armas, L., Ochoa, I. and Lynch, J. P. 2011. Genotypic variation for root traits of maize (*Zea mays* L.)

from the Purhepecha Plateau under contrasting phosphorus availability. *Field Crops Research* 121(3), 350–362.

Becquer, A., Haling, R., Stefanski, A., Richardson, A. and Simpson, R. 2017. Complementary phosphorus acquisition strategies of interplanted subterranean clover and white lupin increase sward yield in a low phosphorus soil. *Doing More with Less*. Proceedings of the 18th Australian Agronomy Conference 2017, Ballarat, Victoria, Australia. Australian Society of Agronomy Inc. pp. 1–4.

Berta, G., Trotta, A., Fusconi, A., Hooker, J. E., Munro, M., Atkinson, D., Giovannetti, M., Morini, S., Fortuna, P., Tisserant, B., Gianinazzi-Pearson, V. and Gianinazzi, S. 1995. Arbuscular mycorrhizal induced changes to plant growth and root system morphology in *Prunus cerasifera*. *Tree Physiology* 15(5), 281–293.

Bishopp, A. and Lynch, J. P. 2015. The hidden half of crop yields. *Nature Plants* 1, 15117.

Blamey, F. P. C., Edwards, D. G. and Asher, C. J. 1983. Effects of aluminum, OH-Al and P-Al molar ratios, and ionic-strength on soybean root elongation in solution culture. *Soil Science* 136(4), 197–207.

Bolland, M. D. A. 1995. *Lupinus cosentinii* more effectively utilizes low levels of phosphorus from superphosphate than *Lupinus angustifolius*. *Journal of Plant Nutrition* 18(3), 421–435.

Bolland, M. D. A. 1997. Comparative phosphorus requirement of four lupin species. *Journal of Plant Nutrition* 20(10), 1239–1253.

Bolland, M. D. A. and Brennan, R. F. 2008. Comparing the phosphorus requirements of wheat, lupin, and canola. *Australian Journal of Agricultural Research* 59(11), 983–998.

Bolland, M. D. A., Siddique, K. H. M., Loss, S. P. and Baker, M. J. 1999. Comparing responses of grain legumes, wheat and canola to applications of superphosphate. *Nutrient Cycling in Agroecosystems* 53(2), 157–175.

Bolland, M. D. A., Sweetingham, M. W. and Jarvis, R. J. 2000. Effect of applied phosphorus on the growth of *Lupinus luteus*, L. *angustifolius* and L. *albus* in acidic soils in the south-west of Western Australia. *Australian Journal of Experimental Agriculture* 40(1), 79–92.

Bovill, W. D., Huang, C. Y. and McDonald, G. K. 2013. Genetic approaches to enhancing phosphorus-use efficiency (PUE) in crops: challenges and directions. *Crop and Pasture Science* 64(3), 179–198.

Brennan, R. F. and Bolland, M. D. A. 2001. Comparing fertilizer phosphorus requirements of canola, lupin, and wheat. *Journal of Plant Nutrition* 24(12), 1885–1900.

Brodelius, P. and Vogel, H. J. 1985. A phosphorus-31 nuclear magnetic resonance study of phosphate uptake and storage in cultured *Catharanthus roseus* and *Daucus carota* plant cells. *Journal of Biological Chemistry* 260(6), 3556–3560.

Brooker, R. W., Bennett, A. E., Cong, W. F., Daniell, T. J., George, T. S., Hallett, P. D., Hawes, C., Iannetta, P. P. M., Jones, H. G., Karley, A. J., Li, L., McKenzie, B. M., Pakeman, R. J., Paterson, E., Schöb, C., Shen, J., Squire, G., Watson, C. A., Zhang, C., Zhang, F., Zhang, J. and White, P. J. 2015. Improving intercropping: a synthesis of research in agronomy, plant physiology and ecology. *New Phytologist* 206(1), 107–117.

Brown, L. K., George, T. S., Barrett, G. E., Hubbard, S. F. and White, P. J. 2013. Interactions between root hair length and arbuscular mycorrhizal colonisation in phosphorus deficient barley (*Hordeum vulgare*). *Plant and Soil* 372(1–2), 195–205.

Brundrett, M. C. 2002. Coevolution of roots and mycorrhizas of land plants. *New Phytologist* 154(2), 275–304.

Brundrett, M. C. 2009. Mycorrhizal associations and other means of nutrition of vascular plants: understanding the global diversity of host plants by resolving conflicting information and developing reliable means of diagnosis. *Plant and Soil* 320(1–2), 37–77.

Bruulsema, T. W., Peterson, H. M. and Prochnow, L. I. 2019. The science of 4R nutrient stewardship for phosphorus management across latitudes. *Journal of Environmental Quality* 48(5), 1295–1299.

Bünemann, E. K., Keller, B., Hoop, D., Jud, K., Boivin, P. and Frossard, E. 2013. Increased availability of phosphorus after drying and rewetting of a grassland soil: processes and plant use. *Plant and Soil* 370(1–2), 511–526.

Calabi-Floody, M., Medina, J., Rumpel, C., Condron, L. M., Hernandez, M., Dumont, M. and Mora, M. D. L. L. 2018. Smart fertilizers as a strategy for sustainable agriculture. *Advances in Agronomy* 147, 119–157.

Caradus, J. R. 2012. Effect of root hair length on white clover growth over a range of soil phosphorus levels. *New Zealand Journal of Agricultural Research* 24(3–4), 353–358.

Chen, W., Koide, R. T., Adams, T. S., Deforest, J. L., Cheng, L. and Eissenstat, D. M. 2016. Root morphology and mycorrhizal symbioses together shape nutrient foraging strategies of temperate trees. *Proceedings of the National Academy of Sciences of the United States of America* 113(31), 8741–8746.

Chen, X., Chen, F., Chen, Y., Gao, Q., Yang, X., Yuan, L., Zhang, F. and Mi, G. 2013. Modern maize hybrids in Northeast China exhibit increased yield potential and resource use efficiency despite adverse climate change. *Global Change Biology* 19(3), 923–936.

Chin, J. H., Gamuyao, R., Dalid, C., Bustamam, M., Prasetiyono, J., Moeljopawiro, S., Wissuwa, M. and Heuer, S. 2011. Developing rice with high yield under phosphorus deficiency: *Pup1* sequence to application. *Plant Physiology* 156(3), 1202–1216.

Chung, H., Zak, D. R., Reich, P. B. and Ellsworth, D. S. 2007. Plant species richness, elevated CO_2, and atmospheric nitrogen deposition alter soil microbial community composition and function. *Global Change Biology* 13(5), 980–989.

Clarholm, M., Skyllberg, U. and Rosling, A. 2015. Organic acid induced release of nutrients from metal-stabilized soil organic matter – the unbutton model. *Soil Biology and Biochemistry* 84, 168–176.

Clark, F. E. 1967. Bacteria in soil. In: Burges, A. and Raw, F. (Eds) *Soil Biology*. London: Academic Press. pp. 15–49.

Cordell, D., Drangert, J. O. and White, S. 2009. The story of phosphorus: global food security and food for thought. *Global Environmental Change* 19(2), 292–305.

Crush, J. R. 1973. The effect of *Rhizophagus tenuis* mycorrhizas on ryegrass, cocksfoot and sweet vernal. *New Phytologist* 72(5), 965–973.

Deng, Y., Chen, K., Teng, W., Zhan, A., Tong, Y., Feng, G., Cui, Z., Zhang, F. and Chen, X. 2014. Is the inherent potential of maize roots efficient for soil phosphorus acquisition? *PLoS ONE* 9(3), e90287.

Deng, Y., Feng, G., Chen, X. and Zou, C. 2017. Arbuscular mycorrhizal fungal colonization is considerable at optimal Olsen-P levels for maximized yields in an intensive wheat-maize cropping system. *Field Crops Research* 209, 1–9.

Dinkelaker, B., Romheld, V. and Marschner, H. 1989. Citric acid excretion and precipitation of calcium citrate in the rhizosphere of white lupin (*Lupinus albus* L.). *Plant, Cell and Environment* 12(3), 285–292.

Dissanayaka, D. M. S., Maruyama, H., Masuda, G. and Wasaki, J. 2015. Interspecific facilitation of P acquisition in intercropping of maize with white lupin in two contrasting soils as influenced by different rates and forms of P supply. *Plant and Soil* 390(1-2), 223-236.

Djigal, D., Brauman, A., Diop, T. A., Chotte, J. L. and Villenave, C. 2004. Influence of bacterial-feeding nematodes (Cephalobidae) on soil microbial communities during maize growth. *Soil Biology and Biochemistry* 36(2), 323-331.

Drew, M. C. 1975. Comparison of the effects of a localised supply of phosphate, nitrate, ammonium and potassium on the growth of the seminal root system, and the shoot, in barley. *New Phytologist* 75(3), 479-490.

Drogue, B., Doré, H., Borland, S., Wisniewski-Dyé, F. and Prigent-Combaret, C. 2012. Which specificity in cooperation between phytostimulating rhizobacteria and plants? *Research in Microbiology* 163(8), 500-510.

Duvick, D. N. 2005. The contribution of breeding to yield advances in maize (*Zea mays* L.). *Advances in Agronomy* 86, 83-145.

Eisenhauer, N. Beßler, H., Engels, C., Gleixner, G., Habekost, M., Milcu, A., Partsch, S., Sabais, A. C. W., Scherber, C., Steinbeiss, S., Weigelt, A., Weisser, W. W. and Scheu, S. 2010. Plant diversity effects on soil microorganisms support the singular hypothesis. *Ecology* 91, 485-496.

Fan, J., Du, Y., Turner, N. C., Wang, B., Fang, Y., Xi, Y., Guo, X. and Li, F. 2015. Changes in root morphology and physiology to limited phosphorus and moisture in a locally-selected cultivar and an introduced cultivar of *Medicago sativa* L. growing in alkaline soil. *Plant and Soil* 392(1-2), 215-226.

Fitter, A. H. 1994. Architecture and biomass allocation as components of the plastic response of root systems to soil heterogeneity. In: Martyn, M. C. and Robert, W. P. (Eds) *Exploitation of Environmental Heterogeneity by Plants*. Boston, MA: Academic Press. pp. 305-323.

Fixen, P. E. and Johnston, A. M. 2012. World fertilizer nutrient reserves: a view to the future. *Journal of the Science of Food and Agriculture* 92(5), 1001-1005.

Freschet, G. T. and Roumet, C. 2017. Sampling roots to capture plant and soil functions. *Functional Ecology* 31(8), 1506-1518.

Frew, A., Powell, J. R., Glauser, G., Bennett, A. E. and Johnson, S. N. 2018. Mycorrhizal fungi enhance nutrient uptake but disarm defences in plant roots, promoting plant-parasitic nematode populations. *Soil Biology and Biochemistry* 126, 123-132.

Furbank, R. T. and Tester, M. 2011. Phenomics – technologies to relieve the phenotyping bottleneck. *Trends in Plant Science* 16(12), 635-644.

Galindo-Castaneda, T., Brown, K. M., Kuldau, G. A., Roth, G. W., Wenner, N. G., Ray, S., Schneider, H. and Lynch, J. P. 2019. Root cortical anatomy is associated with differential pathogenic and symbiotic fungal colonization in maize. *Plant, Cell and Environment* 42(11), 2999-3014.

Gamuyao, R., Chin, J. H., Pariasca-Tanaka, J., Pesaresi, P., Catausan, S., Dalid, C., Slamet-Loedin, I., Tecson-Mendoza, E. M., Wissuwa, M. and Heuer, S. 2012. The protein kinase Pstol1 from traditional rice confers tolerance of phosphorus deficiency. *Nature* 488(7412), 535-539.

Gardner, W. K., Barber, D. A. and Parbery, D. G. 1983a. The acquisition of phosphorus by Lupinus albus L.: III. The probable mechanism by which phosphorus movement in the soil/root interface is enhanced. *Plant and Soil* 70(1), 107-124.

Gardner, W. K. and Boundy, K. A. 1983. The acquisition of phosphorus by *Lupinus albus* L. IV. The effect of interplanting wheat and white lupin on the growth and mineral composition of the two species. *Plant and Soil* 70(3), 391-402.

Gardner, W. K., Parbery, D. G. and Barber, D. A. 1981. Proteoid root morphology and function in *Lupinus albus*. *Plant and Soil* 60(1), 143–147.

Gardner, W. K., Parbery, D. G. and Barber, D. A. 1982a. The acquisition of phosphorus by *Lupinus albus* L. I. Some characteristics of the soil/root interface. *Plant and Soil* 68(1), 19–32.

Gardner, W. K., Parbery, D. G. and Barber, D. A. 1982b. The acquisition of phosphorus by *Lupinus albus* L. II. The effect of varying phosphorus supply and soil type on some characteristics of the soil/root interface. *Plant and Soil* 68(1), 33–41.

Gardner, W. K., Parbery, D. G., Barber, D. A., Swinden, L. J. P. and SOIL 1983b. The acquisition of phosphorus by *Lupinus albus* L. V. The diffusion of exudates away from roots: a computer simulation. *Plant and Soil* 72(1), 13–29.

George, T. S., Quiquampoix, H., Simpson, R. J. and Richardson, A. E. 2007a. Interactions between phytases and soil constituents: implications for the hydrolysis of inositol phosphates. In: Turner, B. L., Richardson, A. E. and Mullaney, E. J. (Eds) *Inositol Phosphates: Linking Agriculture and the Environment*. Wallingford: CAB International. pp. 221–241.

George, T. S., Richardson, A. E., Smith, J. B., Hadobas, P. A. and Simpson, R. J. 2005. Limitations to the potential of transgenic *Trifolium subterraneum* L. plants that exude phytase when grown in soils with a range of organic P content. *Plant and Soil* 278(1–2), 263–274.

George, T. S., Simpson, R. J., Hadobas, P. A., Marshall, D. J. and Richardson, A. E. 2007b. Accumulation and phosphatase-lability of organic phosphorus in fertilised pasture soils. *Australian Journal of Agricultural Research* 58(1), 47–55.

Giles, C. D., Brown, L. K., Adu, M. O., Mezeli, M. M., Sandral, G. A., Simpson, R. J., Wendler, R., Shand, C. A., Menezes-Blackburn, D., Darch, T., Stutter, M. I., Lumsdon, D. G., Zhang, H., Blackwell, M. S., Wearing, C., Cooper, P., Haygarth, P. M. and George, T. S. 2017. Response-based selection of barley cultivars and legume species for complementarity: root morphology and exudation in relation to nutrient source. *Plant Science* 255, 12–28.

Giles, C. D., George, T. S., Brown, L. K., Mezeli, M. M., Richardson, A. E., Shand, C. A., Wendler, R., Darch, T., Menezes-Blackburn, D., Cooper, P., Stutter, M. I., Lumsdon, D. G., Blackwell, M. S. A., Wearing, C., Zhang, H. and Haygarth, P. M. 2016. Does the combination of citrate and phytase exudation in *Nicotiana tabacum* promote the acquisition of endogenous soil organic phosphorus? *Plant and Soil* 412(1–2), 43–59.

Giles, C. D., Richardson, A. E., Cade-Menun, B. J., Mezeli, M. M., Brown, L. K., Menezes-Blackburn, D., Darch, T., Blackwell, M. S., Shand, C. A., Stutter, M. I., Wendler, R., Cooper, P., Lumsdon, D. G., Wearing, C., Zhang, H., Haygarth, P. M. and George, T. S. 2018. Phosphorus acquisition by citrate- and phytase-exuding *Nicotiana tabacum* plant mixtures depends on soil phosphorus availability and root intermingling. *Physiologia Plantarum* 163, 356–371.

Gladstones, J. 1970. Lupins as crop plants. *Field Crop Abstracts* 23, 123–148.

Gong, X., Wheeler, R., Bovill, W. D. and McDonald, G. K. 2016. QTL mapping of grain yield and phosphorus efficiency in barley in a Mediterranean-like environment. *Theoretical and Applied Genetics* 129(9), 1657–1672.

Graham, J. H. and Abbott, L. K. 2000. Wheat responses to aggressive and non-aggressive arbuscular mycorrhizal fungi. *Plant and Soil* 220(1/2), 207–218.

Grant, C. A. and Flaten, D. N. 2019. 4R management of phosphorus fertilizer in the Northern Great Plains. *Journal of Environmental Quality* 48(5), 1356–1369.

Grierson, P. F. 1992. Organic acids in the rhizosphere of *Banksia integrifolia* L.f. *Plant and Soil* 144(2), 259–265.

Haling, R. E., Brown, L. K., Stefanski, A., Kidd, D. R., Ryan, M. H., Sandral, G. A., George, T. S., Lambers, H. and Simpson, R. J. 2018. Differences in nutrient foraging among *Trifolium subterraneum* cultivars deliver improved P-acquisition efficiency. *Plant and Soil* 424(1–2), 539–554.

Haling, R. E., Yang, Z., Shadwell, N., Culvenor, R. A., Stefanski, A., Ryan, M. H., Sandral, G. A., Kidd, D. R., Lambers, H. and Simpson, R. J. 2016. Growth and root dry matter allocation by pasture legumes and a grass with contrasting external critical phosphorus requirements. *Plant and Soil* 407(1–2), 67–79.

Hart, M. M., Antunes, P. M., Chaudhary, V. B., Abbott, L. K. and Field, K. 2018. Fungal inoculants in the field: is the reward greater than the risk? *Functional Ecology* 32(1), 126–135.

Hayes, P., Turner, B. L., Lambers, H. and Laliberté, E. 2014. Foliar nutrient concentrations and resorption efficiency in plants of contrasting nutrient-acquisition strategies along a 2-million-year dune chronosequence. *Journal of Ecology* 102(2), 396–410.

Haynes, R. J. 1990. Active ion uptake and maintenance of cation-anion balance: A critical examination of their role in regulating rhizosphere pH. *Plant and Soil* 126(2), 247–264.

Hetrick, B. A. D., Wilson, G. W. T. and Leslie, J. F. 1991. Root architecture of warm- and cool-season grasses: relationship to mycorrhizal dependence. *Canadian Journal of Botany* 69(1), 112–118.

Hill, J. O., Simpson, R. J., Ryan, M. H. and Chapman, D. F. 2010. Root hair morphology and mycorrhizal colonisation of pasture species in response to phosphorus and nitrogen nutrition. *Crop and Pasture Science* 61(2), 122–131.

Hinsinger, P., Herrmann, L., Lesueur, D., Robin, A., Trap, J., Waithaisong, K. and Plassard, C. 2015. Impact of roots, microorganisms and microfauna on the fate of soil phosphorus in the rhizosphere. *Annual Plant Reviews* 48, 377–408.

Hocking, P. J. and Jeffery, S. 2004. Cluster-root production and organic anion exudation in a group of old-world lupins and a new-world lupin. *Plant and Soil* 258(1), 135–150.

Hodge, A. 2004. The plastic plant: root responses to heterogeneous supplies of nutrients. *New Phytologist* 162(1), 9–24.

Hooper, D. U., Chapin III, F. S., Ewel, J. J., Hector, A., Inchausti, P., Lavorel, S., Lawton, J. H., Lodge, D. M., Loreau, M., Naeem, S., Schmid, B., Setälä, H., Symstad, A. J., Vandermeer, J. and Wardle, D. A. 2005. Effects of biodiversity on ecosystem functioning: A consensus of current knowledge. *Ecological Monographs* 75(1), 3–35.

Hufnagel, B., De Sousa, S. M., Assis, L., Guimaraes, C. T., Leiser, W., Azevedo, G. C., Negri, B., Larson, B. G., Shaff, J. E., Pastina, M. M., Barros, B. A., Weltzien, E., Rattunde, H. F. W., Viana, J. H., Clark, R. T., Falcão, A., Gazaffi, R., Garcia, A. A. F., Schaffert, R. E., Kochian, L. V. and Magalhaes, J. V. 2014. Duplicate and conquer: multiple homologs of *PHOSPHORUS-STARVATION TOLERANCE1* enhance phosphorus acquisition and sorghum performance on low-phosphorus soils. *Plant Physiology* 166(2), 659–677.

Inal, A., Gunes, A., Zhang, F. and Cakmak, I. 2007. Peanut/maize intercropping induced changes in rhizosphere and nutrient concentrations in shoots. *Plant Physiology and Biochemistry* 45(5), 350–356.

Jakobsen, I., Chen, B., Munkvold, L., Lundsgaard, T. and Zhu, Y.-G. 2005. Contrasting phosphate acquisition of mycorrhizal fungi with that of root hairs using the root hairless barley mutant. *Plant, Cell and Environment* 28(7), 928–938.

Jeffery, R. P., Simpson, R. J., Lambers, H., Kidd, D. R. and Ryan, M. H. 2017a. Plants in constrained canopy micro-swards compensate for decreased root biomass and soil exploration with increased amounts of rhizosphere carboxylates. *Functional Plant Biology* 44(5), 552–562.

Jeffery, R. P., Simpson, R. J., Lambers, H., Kidd, D. R. and Ryan, M. H. 2017b. Root morphology acclimation to phosphorus supply by six cultivars of *Trifolium subterraneum* L. *Plant and Soil* 412(1–2), 21–34.

Jeffery, R. P., Simpson, R. J., Lambers, H., Orchard, S., Kidd, D. R., Haling, R. E. and Ryan, M. H. 2018. Contrasting communities of arbuscule-forming root symbionts change external critical phosphorus requirements of some annual pasture legumes. *Applied Soil Ecology* 126, 88–97.

Jemo, M., Abaidoo, R. C., Nolte, C., Tchienkoua, M., Sanginga, N. and Horst, W. J. 2006. Phosphorus benefits from grain-legume crops to subsequent maize grown on acid soils of southern Cameroon. *Plant and Soil* 284(1–2), 385–397.

Jia, X., Liu, P. and Lynch, J. P. 2018. Greater lateral root branching density in maize improves phosphorus acquisition from low phosphorus soil. *Journal of Experimental Botany* 69(20), 4961–4970.

Johnson, N. C., Graham, J. H. and Smith, F. A. 1997. Functioning of mycorrhizal associations along the mutualism–parasitism continuum. *New Phytologist* 135(4), 575–585.

Jones, D. L. 1998. Organic acids in the rhizosphere. *Plant and Soil* 205(1), 25–44.

Jones, D. L., Dennis, P. G., Owen, A. G. and Van Hees, P. A. W. 2003. Organic acid behavior in soils - misconceptions and knowledge gaps. *Plant and Soil* 248(1/2), 31–41.

Keerthisinghe, G., Hocking, P. J., Ryan, P. R. and Delhaize, E. 1998. Effect of phosphorus supply on the formation and function of proteoid roots of white lupin (*Lupinus albus* L.). *Plant, Cell and Environment* 21(5), 467–478.

Khan, M. S., Zaidi, A. and Musarrad, J. 2014. *Phosphate Solubilizing Microorganisms, Principles and Application of Microphos Technology*. New Delhi: Springer International Publishing.

Kidd, D. R., Ryan, M. H., Hahne, D., Haling, R. E., Lambers, H., Sandral, G. A., Simpson, R. J. and Cawthray, G. R. 2018. The carboxylate composition of rhizosheath and root exudates from twelve species of grassland and crop legumes with special reference to the occurrence of citramalate. *Plant and Soil* 424(1–2), 389–403.

Kidd, D. R., Ryan, M. H., Haling, R. E., Lambers, H., Sandral, G. A., Yang, Z., Culvenor, R. A., Cawthray, G. R., Stefanski, A. and Simpson, R. J. 2015. Rhizosphere carboxylates and morphological root traits in pasture legumes and grasses. *Plant and Soil* 402(1–2), 77–89.

Kirkby, E. A. and Mengel, K. 1967. Ionic balance in different tissues of the tomato plant in relation to nitrate, urea, or ammonium nutrition. *Plant Physiology* 42(1), 6–14.

Kong, Y., Li, X., Ma, J., Li, W., Yan, G. and Zhang, C. 2014. *GmPAP4*, a novel purple acid phosphatase gene isolated from soybean (*Glycine max*), enhanced extracellular phytate utilization in *Arabidopsis thaliana*. *Plant Cell Reports* 33(4), 655–667.

Kuzyakov, Y. and Razavi, B. S. 2019. Rhizosphere size and shape: temporal dynamics and spatial stationarity. *Soil Biology and Biochemistry* 135, 343–360.

Lambers, H., Albornoz, F., Kotula, L., Laliberté, E., Ranathunge, K., Teste, F. P. and Zemunik, G. 2018. How belowground interactions contribute to the coexistence of mycorrhizal and non-mycorrhizal species in severely phosphorus-impoverished hyperdiverse ecosystems. *Plant and Soil* 424(1–2), 11–33.

Lambers, H., Clements, J. C. and Nelson, M. N. 2013. How a phosphorus-acquisition strategy based on carboxylate exudation powers the success and agronomic potential of lupines (*Lupinus*, Fabaceae). *American Journal of Botany* 100(2), 263-288.

Lambers, H., Finnegan, P. M., Laliberte, E., Pearse, S. J., Ryan, M. H., Shane, M. W. and Veneklaas, E. J. 2011. Update on phosphorus nutrition in Proteaceae. Phosphorus nutrition of Proteaceae in severely phosphorus-impoverished soils: are there lessons to be learned for future crops? *Plant Physiology* 156(3), 1058-1066.

Lambers, H., Hayes, P. E., Laliberté, E., Oliveira, R. S. and Turner, B. L. 2015a. Leaf manganese accumulation and phosphorus-acquisition efficiency. *Trends in Plant Science* 20(2), 83-90.

Lambers, H., Martinoia, E. and Renton, M. 2015b. Plant adaptations to severely phosphorus-impoverished soils. *Current Opinion in Plant Biology* 25, 23-31.

Lambers, H., Raven, J. A., Shaver, G. R. and Smith, S. E. 2008. Plant nutrient-acquisition strategies change with soil age. *Trends in Ecology and Evolution* 23(2), 95-103.

Lambers, H., Shane, M. W., Cramer, M. D., Pearse, S. J. and Veneklaas, E. J. 2006. Root structure and functioning for efficient acquisition of phosphorus: matching morphological and physiological traits. *Annals of Botany* 98(4), 693-713.

Lee, R. B. and Ratcliffe, R. G. 1993. Subcellular distribution of inorganic phosphate, and levels of nucleoside triphosphate, in mature maize roots at low external phosphate concentrations: measurements with ^{31}P-NMR. *Journal of Experimental Botany* 44(3), 587-598.

Leiser, W. L., Rattunde, H. F. W., Weltzien, E., Cisse, N., Abdou, M., Diallo, A., Tourè, A. O., Magalhaes, J. V. and Haussmann, B. I. G. 2014. Two in one sweep: aluminum tolerance and grain yield in P-limited soils are associated to the same genomic region in West African sorghum. *BMC Plant Biology* 14, 206-206.

Li, H., Huang, G., Meng, Q., Ma, L., Yuan, L., Wang, F., Zhang, W., Cui, Z., Shen, J., Chen, X., Jiang, R. and Zhang, F. 2011. Integrated soil and plant phosphorus management for crop and environment in China: a review. *Plant and Soil* 349(1-2), 157-167.

Li, H., Liu, B., Mccormack, M. L., Ma, Z. and Guo, D. 2017. Diverse belowground resource strategies underlie plant species coexistence and spatial distribution in three grasslands along a precipitation gradient. *New Phytologist* 216(4), 1140-1150.

Li, H., Shen, J., Zhang, F., Clairotte, M., Drevon, J. J., Le Cadre, E. and Hinsinger, P. 2008. Dynamics of phosphorus fractions in the rhizosphere of common bean (*Phaseolus vulgaris* L.) and durum wheat (*Triticum turgidum durum* L.) grown in monocropping and intercropping systems. *Plant and Soil* 312(1-2), 139-150.

Li, H., Shen, J., Zhang, F., Marschner, P., Cawthray, G. and Rengel, Z. 2010. Phosphorus uptake and rhizosphere properties of intercropped and monocropped maize, faba bean, and white lupin in acidic soil. *Biology and Fertility of Soils* 46(2), 79-91.

Li, H., Ma, Q., Li, H., Zhang, F., Rengel, Z. and Shen, J. 2014a. Root morphological responses to localized nutrient supply differ among crop species with contrasting root traits. *Plant and Soil* 376(1-2), 151-163.

Li, L., Li, S. M., Sun, J. H., Zhou, L. L., Bao, X. G., Zhang, H. G. and Zhang, F. S. 2007. Diversity enhances agricultural productivity via rhizosphere phosphorus facilitation on phosphorus-deficient soils. *Proceedings of the National Academy of Sciences of the United States of America* 104(27), 11192-11196.

Li, L., Tang, C., Rengel, Z. and Zhang, F. 2003. Chickpea facilitates phosphorus uptake by intercropped wheat from an organic phosphorus source. *Plant and Soil* 248(1/2), 297-303.

Li, L., Tilman, D., Lambers, H. and Zhang, F. S. 2014b. Plant diversity and overyielding: insights from belowground facilitation of intercropping in agriculture. *New Phytologist* 203(1), 63–69.

Li, L., Yang, S., Li, X., Zhang, F. and Christie, P. 1999. Interspecific complementary and competitive interactions between intercropped maize and faba bean. *Plant and Soil* 212(2), 105–114.

Li, R., Lu, W., Guo, C., Li, X., Gu, J. and Xiao, K. 2012. Molecular characterization and functional analysis of OsPHY1, a purple acid phosphatase (PAP)-type phytase gene in rice (*Oryza sativa* L.). *Journal of Integrative Agriculture* 11(8), 1217–1226.

Li, S. M., Li, L., Zhang, F. S. and Tang, C. 2004. Acid phosphatase role in chickpea/maize intercropping. *Annals of Botany* 94(2), 297–303.

Lipton, D. S., Blanchar, R. W. and Blevins, D. G. 1987. Citrate, malate, and succinate concentration in exudates from P-sufficient and P-stressed *Medicago sativa* L. seedlings. *Plant Physiology* 85(2), 315–317.

Liu, B., Li, H., Zhu, B., Koide, R. T., Eissenstat, D. M. and Guo, D. 2015. Complementarity in nutrient foraging strategies of absorptive fine roots and arbuscular mycorrhizal fungi across 14 coexisting subtropical tree species. *New Phytologist* 208(1), 125–136.

Liu, H. T., White, P. J. and Li, C. J. 2016a. Biomass partitioning and rhizosphere responses of maize and faba bean to phosphorus deficiency. *Crop and Pasture Science* 67(8), 847–856.

Liu, T. Y., Huang, T. K., Yang, S. Y., Hong, Y. T., Huang, S. M., Wang, F. N., Chiang, S. F., Tsai, S. Y., Lu, W. C. and Chiou, T. J. 2016b. Identification of plant vacuolar transporters mediating phosphate storage. *Nature Communications* 7, 11095.

Liu, Z., Gao, K., Shan, S., Gu, R., Wang, Z., Craft, E. J., Mi, G., Yuan, L. and Chen, F. 2017. Comparative analysis of root traits and the associated QTLs for maize seedlings grown in paper roll, hydroponics and vermiculite culture system. *Frontiers in Plant Science* 8, 436.

Lott, J. N., Greenwood, J. S., Vollmer, C. M. and Buttrose, M. S. 1978. Energy-dispersive X-ray analysis of phosphorus, potassium, magnesium, and calcium in globoid crystals in protein bodies from different regions of *Cucurbita maxima* embryos. *Plant Physiology* 61(6), 984–988.

Lynch, J. P. 2015. Root phenes that reduce the metabolic costs of soil exploration: opportunities for 21st century agriculture. *Plant, Cell and Environment* 38(9), 1775–1784.

Lynch, J. P. 2019. Root phenotypes for improved nutrient capture: an underexploited opportunity for global agriculture. *New Phytologist* 223(2), 548–564.

Lynch, J. P. and Brown, K. M. 2008. Root strategies for phosphorus acquisition. In: White, P. J. and Hammond, J. P. (Eds) *The Ecophysiology of Plant-Phosphorus Interactions*. Netherlands: Springer. pp. 83–116.

Lynch, J. P. and Ho, M. D. 2005. Rhizoeconomics: carbon costs of phosphorus acquisition. *Plant and Soil* 269(1–2), 45–56.

Lyu, Y., Tang, H., Li, H., Zhang, F., Rengel, Z., Whalley, W. R. and Shen, J. 2016. Major crop species show differential balance between root morphological and physiological responses to variable phosphorus supply. *Frontiers in Plant Science* 7, 1939.

Ma, X. F., Tudor, S., Butler, T., Ge, Y., Xi, Y., Bouton, J., Harrison, M. and Wang, Z. Y. 2012. Transgenic expression of phytase and acid phosphatase genes in alfalfa (*Medicago sativa*) leads to improved phosphate uptake in natural soils. *Molecular Breeding* 30(1), 377–391.

Ma, X. M., Mason-Jones, K., Liu, Y., Blagodatskaya, E., Kuzyakov, Y., Guber, A., Dippold, M. A. and Razavi, B. S. 2019. Coupling zymography with pH mapping reveals a shift in lupine phosphorus acquisition strategy driven by cluster roots. *Soil Biology and Biochemistry* 135, 420–428.

Ma, Z., Guo, D., Xu, X., Lu, M., Bardgett, R. D., Eissenstat, D. M., McCormack, M. L. and Hedin, L. O. 2018. Evolutionary history resolves global organization of root functional traits. *Nature* 555(7694), 94–97.

MacDonald, G. K., Bennett, E. M., Potter, P. A. and Ramankutty, N. 2011. Agronomic phosphorus imbalances across the world's croplands. *Proceedings of the National Academy of Sciences of the United States of America* 108(7), 3086–3091.

Maherali, H. 2014. Is there an association between root architecture and mycorrhizal growth response? *New Phytologist* 204(1), 192–200.

Mai, W., Xue, X., Feng, G., Yang, R. and Tian, C. 2018. Can optimization of phosphorus input lead to high productivity and high phosphorus use efficiency of cotton through maximization of root/mycorrhizal efficiency in phosphorus acquisition? *Field Crops Research* 216, 100–108.

Marschner, H., Romheld, V. and Cakmak, I. 1987. Root-induced changes of nutrient availability in the rhizosphere. *Journal of Plant Nutrition* 10(9), 1175–1184.

Martin-Robles, N., Lehmann, A., Seco, E., Aroca, R., Rillig, M. C. and Milla, R. 2018. Impacts of domestication on the arbuscular mycorrhizal symbiosis of 27 crop species. *New Phytologist* 218(1), 322–334.

Martin, B. C., George, S. J., Price, C. A., Shahsavari, E., Ball, A. S., Tibbett, M. and Ryan, M. H. 2016. Citrate and malonate increase microbial activity and alter microbial community composition in uncontaminated and diesel-contaminated soil microcosms. *SOIL* 2(3), 487–498.

Maruyama, H. and Wasaki, J. 2017. Transgenic approaches for improving phosphorus use efficiency in plants. In: Hossain, M. A., Kamiya, T., Burritt, D. J., Tran, L.-S. P. and Fujiwara, T. (Eds) *Plant Macronutrient Use Efficiency*. London: Academic Press. pp. 323–338.

McCormack, M. L., Dickie, I. A., Eissenstat, D. M., Fahey, T. J., Fernandez, C. W., Guo, D., Helmisaari, H. S., Hobbie, E. A., Iversen, C. M., Jackson, R. B., Leppalammi-Kujansuu, J., Norby, R. J., Phillips, R. P., Pregitzer, K. S., Pritchard, S. G., Rewald, B. and Zadworny, M. 2015. Redefining fine roots improves understanding of below-ground contributions to terrestrial biosphere processes. *New Phytologist* 207(3), 505–518.

McLaughlin, M. J., Mcbeath, T. M., Smernik, R., Stacey, S. P., Ajiboye, B. and Guppy, C. 2011. The chemical nature of P accumulation in agricultural soils—implications for fertiliser management and design: an Australian perspective. *Plant and Soil* 349(1–2), 69–87.

Mendoza, R. 2001. Phosphorus nutrition and mycorrhizal growth response of broadleaf and narrowleaf birdsfoot trefoils. *Journal of Plant Nutrition* 24(1), 203–214.

Menezes-Blackburn, D., Giles, C., Darch, T., George, T. S., Blackwell, M., Stutter, M., Shand, C., Lumsdon, D., Cooper, P., Wendler, R., Brown, L., Almeida, D. S., Wearing, C., Zhang, H. and Haygarth, P. M. 2018. Opportunities for mobilizing recalcitrant phosphorus from agricultural soils: a review. *Plant and Soil* 427(1), 5–16.

Menezes-Blackburn, D., Jorquera, M. A., Greiner, R., Gianfreda, L. and De La Luz Mora, M. J. C. R. I. E. S. 2013. Phytases and phytase-labile organic phosphorus in manures and soils. *Critical Reviews in Environmental Science and Technology* 43(9), 916–954.

Mi, G., Chen, F., Yuan, L. and Zhang, F. 2016. Ideotype root system architecture for maize to achieve high yield and resource use efficiency in intensive cropping systems. *Advances in Agronomy* 139, 73-97.

Miguel, M. A., Postma, J. A. and Lynch, J. P. 2015. Phene synergism between root hair length and basal root growth angle for phosphorus acquisition. *Plant Physiology* 167(4), 1430-1439.

Muler, A. L., Oliveira, R. S., Lambers, H. and Veneklaas, E. J. 2014. Does cluster-root activity benefit nutrient uptake and growth of co-existing species? *Oecologia* 174(1), 23-31.

Nadeem, M., Mollier, A., Morel, C., Vives, A., Prud'homme, L. and Pellerin, S. 2011. Relative contribution of seed phosphorus reserves and exogenous phosphorus uptake to maize (*Zea mays* L.) nutrition during early growth stages. *Plant and Soil* 346(1-2), 231-244.

Nadeem, M., Mollier, A., Morel, C., Vives, A., Prud'homme, L. and Pellerin, S. 2012. Maize (*Zea mays* L.) endogenous seed phosphorus remobilization is not influenced by exogenous phosphorus availability during germination and early growth stages. *Plant and Soil* 357(1-2), 13-24.

Nadeem, S. M., Ahmad, M., Zahir, Z. A., Javaid, A. and Ashraf, M. 2014. The role of mycorrhizae and plant growth promoting rhizobacteria (PGPR) in improving crop productivity under stressful environments. *Biotechnology Advances* 32(2), 429-448.

Nazeri, N. K., Lambers, H., Tibbett, M. and Ryan, M. H. 2014. Moderating mycorrhizas: arbuscular mycorrhizas modify rhizosphere chemistry and maintain plant phosphorus status within narrow boundaries. *Plant, Cell and Environment* 37(4), 911-921.

Ning, L., Kan, G., Du, W., Guo, S., Wang, Q., Zhang, G., Cheng, H. and Yu, D. 2016. Association analysis for detecting significant single nucleotide polymorphisms for phosphorus-deficiency tolerance at the seedling stage in soybean [*Glycine max* (L) Merr.]. *Breeding Science* 66(2), 191-203.

Nuruzzaman, M., Lambers, H., Bolland, M. D. and Veneklaas, E. J. 2005a. Phosphorus benefits of different legume crops to subsequent wheat grown in different soils of Western Australia. *Plant and Soil* 271(1-2), 175-187.

Nuruzzaman, M., Lambers, H., Bolland, M. D. A. and Veneklaas, E. J. 2005b. Phosphorus uptake by grain legumes and subsequently grown wheat at different levels of residual phosphorus fertiliser. *Australian Journal of Agricultural Research* 56(10), 1041-1047.

Oberson, A. and Joner, E. J. 2005. Microbial turnover of phosphorus in soil. In: Turner, B., Frossard, E. and Baldwin, D. (Eds) *Organic Phosphorus in the Environment*. Wallingford: CABI Publishing. pp. 133-164.

Oburger, E., Dell'Mour, M., Hann, S., Wieshammer, G., Puschenreiter, M. and Wenzel, W. W. 2013. Evaluation of a novel tool for sampling root exudates from soil-grown plants compared to conventional techniques. *Environmental and Experimental Botany* 87, 235-247.

Oburger, E. and Jones, D. L. 2018. Sampling root exudates–mission impossible? *Rhizosphere* 6, 116-133.

Oburger, E., Kirk, G. J. D., Wenzel, W. W., Puschenreiter, M. and Jones, D. L. 2009. Interactive effects of organic acids in the rhizosphere. *Soil Biology and Biochemistry* 41(3), 449-457.

Orchard, S., Hilton, S., Bending, G. D., Dickie, I. A., Standish, R. J., Gleeson, D. B., Jeffery, R. P., Powell, J. R., Walker, C., Bass, D., Monk, J., Simonin, A. and Ryan, M. H. 2017a. Fine endophytes (*Glomus tenue*) are related to Mucoromycotina, not Glomeromycota. *New Phytologist* 213(2), 481–486.

Orchard, S., Standish, R. J., Dickie, I. A., Renton, M., Walker, C., Moot, D. and Ryan, M. H. 2017b. Fine root endophytes under scrutiny: a review of the literature on arbuscule-producing fungi recently suggested to belong to the Mucoromycotina. *Mycorrhiza* 27(7), 619–638.

Pang, J., Bansal, R., Zhao, H., Bohuon, E., Lambers, H., Ryan, M. H., Ranathunge, K. and Siddique, K. H. M. 2018. The carboxylate-releasing phosphorus-mobilizing strategy can be proxied by foliar manganese concentration in a large set of chickpea germplasm under low phosphorus supply. *New Phytologist* 219(2), 518–529.

Parfitt, R. L. 1979. The availability of P from phosphate-goethite bridging complexes. Desorption and uptake by ryegrass. *Plant and Soil* 53(1–2), 55–65.

Pearse, S. J., Veneklaas, E. J., Cawthray, G., Bolland, M. D. and Lambers, H. 2007. Carboxylate composition of root exudates does not relate consistently to a crop species' ability to use phosphorus from aluminium, iron or calcium phosphate sources. *New Phytologist* 173(1), 181–190.

Pearse, S. J., Veneklaas, E. J., Cawthray, G. R., Bolland, M. D. A. and Lambers, H. 2006. Carboxylate release of wheat, canola and 11 grain legume species as affected by phosphorus status. *Plant and Soil* 288(1–2), 127–139.

Postma, J. A., Dathe, A. and Lynch, J. P. 2014. The optimal lateral root branching density for maize depends on nitrogen and phosphorus availability. *Plant Physiology* 166(2), 590–602.

Postma, J. A. and Lynch, J. P. 2011. Root cortical aerenchyma enhances the growth of maize on soils with suboptimal availability of nitrogen, phosphorus, and potassium. *Plant Physiology* 156(3), 1190–1201.

Raboy, V. 2009. Approaches and challenges to engineering seed phytate and total phosphorus. *Plant Science* 177(4), 281–296.

Raven, J. A., Lambers, H., Smith, S. E. and Westoby, M. 2018. Costs of acquiring phosphorus by vascular land plants: patterns and implications for plant coexistence. *New Phytologist* 217(4), 1420–1427.

Razavi, B. S., Zarebanadkouki, M., Blagodatskaya, E. and Kuzyakov, Y. 2016. Rhizosphere shape of lentil and maize: spatial distribution of enzyme activities. *Soil Biology and Biochemistry* 96, 229–237.

Razavi, B. S., Zhang, X., Bilyera, N., Guber, A. and Zarebanadkouki, M. 2019. Soil zymography: simple and reliable? Review of current knowledge and optimization of the method. *Rhizosphere* 11, 100161.

Reinhart, K. O., Wilson, G. W. and Rinella, M. J. 2012. Predicting plant responses to mycorrhizae: integrating evolutionary history and plant traits. *Ecology Letters* 15(7), 689–695.

Richardson, A. E., Lynch, J. P., Ryan, P. R., Delhaize, E., Smith, F. A., Smith, S. E., Harvey, P. R., Ryan, M. H., Veneklaas, E. J., Lambers, H., Oberson, A., Culvenor, R. A. and Simpson, R. J. 2011. Plant and microbial strategies to improve the phosphorus efficiency of agriculture. *Plant and Soil* 349(1–2), 121–156.

Rillig, M. C., Aguilar-Trigueros, C. A., Camenzind, T., Cavagnaro, T. R., Degrune, F., Hohmann, P., Lammel, D. R., Mansour, I., Roy, J., Van Der Heijden, M. G. A. and Yang, G. 2019. Why farmers should manage the arbuscular mycorrhizal symbiosis. *New Phytologist* 222(3), 1171–1175.

Roumet, C., Birouste, M., Picon-Cochard, C., Ghestem, M., Osman, N., Vrignon-Brenas, S., Cao, K. F. and Stokes, A. 2016. Root structure-function relationships in 74 species: evidence of a root economics spectrum related to carbon economy. *New Phytologist* 210(3), 815-826.

Rózycki, H. and Strzelczyk, E. 1986. Organic acids production by *Streptomyces* spp. isolated from soil, rhizosphere and mycorrhizosphere of pine (*Pinus sylvestris* L.). *Plant and Soil* 96(3), 337-345.

Ryan, M. H. and Graham, J. H. 2018. Little evidence that farmers should consider abundance or diversity of arbuscular mycorrhizal fungi when managing crops. *New Phytologist* 220(4), 1092-1107.

Ryan, M. H., Graham, J. H., Morton, J. B. and Kirkegaard, J. A. 2019a. Research must use a systems agronomy approach if management of the arbuscular mycorrhizal symbiosis is to contribute to sustainable intensification. *New Phytologist* 222(3), 1176-1178.

Ryan, M. H., Herwaarden, A. F. V., Angus, J. F. and Kirkegaard, J. A. 2005. Reduced growth of autumn-sown wheat in a low-P soil is associated with high colonisation by arbuscular mycorrhizal fungi. *Plant and Soil* 270(1), 275-286.

Ryan, M. H., Kaur, P., Nazeri, N. K., Clode, P. L., Keeble-Gagnere, G., Doolette, A. L., Smernik, R. J., Van Aken, O., Nicol, D., Maruyama, H., Ezawa, T., Lambers, H., Millar, A. H. and Appels, R. 2019b. Globular structures in roots accumulate phosphorus to extremely high concentrations following phosphorus addition. *Plant, Cell and Environment* 42(6), 1987-2002.

Ryan, M. H., Kidd, D. R., Sandral, G. A., Yang, Z., Lambers, H., Culvenor, R. A., Stefanski, A., Nichols, P. G. H., Haling, R. E. and Simpson, R. J. 2016. High variation in the percentage of root length colonised by arbuscular mycorrhizal fungi among 139 lines representing the species subterranean clover (*Trifolium subterraneum*). *Applied Soil Ecology* 98, 221-232.

Ryan, M. H., Tibbett, M., Edmonds-Tibbett, T., Suriyagoda, L. D., Lambers, H., Cawthray, G. R. and Pang, J. 2012. Carbon trading for phosphorus gain: the balance between rhizosphere carboxylates and arbuscular mycorrhizal symbiosis in plant phosphorus acquisition. *Plant, Cell and Environment* 35(12), 2170-2180.

Ryan, P. R., James, R. A., Weligama, C., Delhaize, E., Rattey, A., Lewis, D. C., Bovill, W. D., Mcdonald, G., Rathjen, T. M., Wang, E., Fettell, N. A. and Richardson, A. E. 2014. Can citrate efflux from roots improve phosphorus uptake by plants? Testing the hypothesis with near-isogenic lines of wheat. *Physiologia Plantarum* 151(3), 230-242.

Salvioli, A., Ghignone, S., Novero, M., Navazio, L., Venice, F., Bagnaresi, P. and Bonfante, P. 2016. Symbiosis with an endobacterium increases the fitness of a mycorrhizal fungus, raising its bioenergetic potential. *The ISME Journal* 10(1), 130-144.

Sattari, S. Z., Bouwman, A. F., Giller, K. E. and Van Ittersum, M. K. 2012. Residual soil phosphorus as the missing piece in the global phosphorus crisis puzzle. *Proceedings of the National Academy of Sciences of the United States of America* 109(16), 6348-6353.

Schefe, C. R., Watt, M., Slattery, W. J. and Mele, P. M. 2008. Organic anions in the rhizosphere of Al-tolerant and Al-sensitive wheat lines grown in an acid soil in controlled and field environments. *Soil Research* 46(3), 257-264.

Schelud'ko, A. V., Makrushin, K. V., Tugarova, A. V., Krestinenko, V. A., Panasenko, V. I., Antonyuk, L. P. and Katsy, E. I. 2009. Changes in motility of the rhizobacterium *Azospirillum brasilense* in the presence of plant lectins. *Microbiological Research* 164(2), 149-156.

Schneider, H. M. and Lynch, J. P. 2018. Functional implications of root cortical senescence for soil resource capture. *Plant and Soil* 423(1–2), 13–26.

Shane, M. W., Cawthray, G. R., Cramer, M. D., Kuo, J. and Lambers, H. 2006. Specialized 'dauciform' roots of Cyperaceae are structurally distinct, but functionally analogous with 'cluster' roots. *Plant, Cell and Environment* 29(10), 1989–1999.

Shane, M. W., De Vos, M., De Roock, S. and Lambers, H. 2003. Shoot P status regulates cluster-root growth and citrate exudation in Lupinus albus grown with a divided root system. *Plant, Cell and Environment* 26(2), 265–273.

Shane, M. W. and Lambers, H. 2005. Cluster roots: a curiosity in context. *Plant and Soil* 274(1–2), 101–125.

Shao, H., Shi, D. F., Shi, W. J., Ban, X. B., Chen, Y. C., Ren, W., Chen, F. J. and Mi, G. H. 2019. Genotypic difference in the plasticity of root system architecture of field-grown maize in response to plant density. *Plant and Soil* 439(1–2), 201–217.

Shao, H., Xia, T., Wu, D., Chen, F. and Mi, G. 2018. Root growth and root system architecture of field-grown maize in response to high planting density. *Plant and Soil* 430(1–2), 395–411.

Shen, J., Li, C., Mi, G., Li, L., Yuan, L., Jiang, R. and Zhang, F. 2013. Maximizing root/rhizosphere efficiency to improve crop productivity and nutrient use efficiency in intensive agriculture of China. *Journal of Experimental Botany* 64(5), 1181–1192.

Shen, J., Rengel, Z., Tang, C. and Zhang, F. 2003. Role of phosphorus nutrition in development of cluster roots and release of carboxylates in soil-grown *Lupinus albus*. *Plant and Soil* 248(1/2), 199–206.

Shen, J., Yuan, L., Zhang, J., Li, H., Bai, Z., Chen, X., Zhang, W. and Zhang, F. 2011. Phosphorus dynamics: from soil to plant. *Plant Physiology* 156(3), 997–1005.

Simpson, R. J., Oberson, A., Culvenor, R. A., Ryan, M. H., Veneklaas, E. J., Lambers, H., Lynch, J. P., Ryan, P. R., Delhaize, E., Smith, F. A., Smith, S. E., Harvey, P. R. and Richardson, A. E. 2011. Strategies and agronomic interventions to improve the phosphorus-use efficiency of farming systems. *Plant and Soil* 349(1–2), 89–120.

Simpson, R. J., Stefanski, A., Marshall, D. J., Moore, A. D. and Richardson, A. E. 2015. Management of soil phosphorus fertility determines the phosphorus budget of a temperate grazing system and is the key to improving phosphorus efficiency. *Agriculture, Ecosystems and Environment* 212, 263–277.

Singh, N., Kumar, S., Bajpai, V. K., Dubey, R. C., Maheshwari, D. K. and Kang, S. C. 2010. Biological control of *Macrophomina phaseolina* by chemotactic fluorescent *Pseudomonas aeruginosa* PN1 and its plant growth promotory activity in chir-pine. *Crop Protection* 29(10), 1142–1147.

Smith, S. and Read, D. (Eds.) 2008. *Mycorrhizal Symbiosis*. London: Academic Press.

Song, Y. N., Zhang, F. S., Marschner, P., Fan, F. L., Gao, H. M., Bao, X. G., Sun, J. H. and Li, L. 2007. Effect of intercropping on crop yield and chemical and microbiological properties in rhizosphere of wheat (*Triticum aestivum* L.), maize (*Zea mays* L.), and faba bean (*Vicia faba* L.). *Biology and Fertility of Soils* 43(5), 565–574.

Spohn, M. and Kuzyakov, Y. 2013. Distribution of microbial- and root-derived phosphatase activities in the rhizosphere depending on P availability and C allocation – coupling soil zymography with ^{14}C imaging. *Soil Biology and Biochemistry* 67, 106–113.

Spohn, M., Treichel, N. S., Cormann, M., Schloter, M. and Fischer, D. 2015. Distribution of phosphatase activity and various bacterial phyla in the rhizosphere of Hordeum vulgare L. depending on P availability. *Soil Biology and Biochemistry* 89, 44–51.

Strock, C. F., Morrow De La Riva, L. and Lynch, J. P. 2018. Reduction in root secondary growth as a strategy for phosphorus acquisition. *Plant Physiology* 176(1), 691–703.

Sun, B., Gao, Y. and Lynch, J. P. 2018. Large crown root number improves topsoil foraging and phosphorus acquisition. *Plant Physiology* 177(1), 90–104.

Sun, X. and Tang, M. 2013. Effect of arbuscular mycorrhizal fungi inoculation on root traits and root volatile organic compound emissions of *Sorghum bicolor*. *South African Journal of Botany* 88, 373–379.

Sungthongwises, K. 2016. Diversity of phosphate solubilizing bacteria under rubber intercropping. *Asian Journal of Plant Sciences* 15(3), 75–80.

Syer, J., Johnston, A. and Curtin, D. 2008. Efficiency of soil and fertilizer phosphorus use – reconciling changing concepts of soil phosphorus behaviour with agronomic information. FAO Fertilizer and Plant Nutrition Bulletin 18. Rome.

Tang, X., Bernard, L., Brauman, A., Daufresne, T., Deleporte, P., Desclaux, D., Souche, G., Placella, S. A. and Hinsinger, P. 2014. Increase in microbial biomass and phosphorus availability in the rhizosphere of intercropped cereal and legumes under field conditions. *Soil Biology and Biochemistry* 75, 86–93.

Tarafdar, J. C. and Jungk, A. 1987. Phosphatase activity in the rhizosphere and its relation to the depletion of soil organic phosphorus. *Biology and Fertility of Soils* 3(4), 199–204.

Taurian, T., Anzuay, M. S., Angelini, J. G., Tonelli, M. L., Ludueña, L., Pena, D., Ibáñez, F. and Fabra, A. 2010. Phosphate-solubilizing peanut associated bacteria: screening for plant growth-promoting activities. *Plant and Soil* 329(1–2), 421–431.

Teng, W., Deng, Y., Chen, X. P., Xu, X. F., Chen, R. Y., Lv, Y., Zhao, Y. Y., Zhao, X. Q., He, X., Li, B., Tong, Y. P., Zhang, F. S. and Li, Z. S. 2013. Characterization of root response to phosphorus supply from morphology to gene analysis in field-grown wheat. *Journal of Experimental Botany* 64(5), 1403–1411.

Teste, F. P., Veneklaas, E. J., Dixon, K. W. and Lambers, H. 2014. Complementary plant nutrient-acquisition strategies promote growth of neighbour species. *Functional Ecology* 28(4), 819–828.

Ubbens, J. R. and Stavness, I. 2017. Deep plant phenomics: a deep learning platform for complex plant phenotyping tasks. *Frontiers in Plant Science* 8, 1190.

Valenzuela-Estrada, L. R., Vera-Caraballo, V., Ruth, L. E. and Eissenstat, D. M. 2008. Root anatomy, morphology, and longevity among root orders in *Vaccinium corymbosum* (Ericaceae). *American Journal of Botany* 95(12), 1506–1514.

Van De Wiel, C. C. M., Van Der Linden, C. G. and Scholten, O. E. 2016. Improving phosphorus use efficiency in agriculture: opportunities for breeding. *Euphytica* 207(1), 1–22.

Vance, C. P., Uhde-Stone, C. and Allan, D. L. 2003. Phosphorus acquisition and use: critical adaptations by plants for securing a nonrenewable resource. *New Phytologist* 157(3), 423–447.

Veneklaas, E. J., Lambers, H., Bragg, J., Finnegan, P. M., Lovelock, C. E., Plaxton, W. C., Price, C. A., Scheible, W. R., Shane, M. W., White, P. J. and Raven, J. A. 2012. Opportunities for improving phosphorus-use efficiency in crop plants. *New Phytologist* 195(2), 306–320.

Veneklaas, E. J., Stevens, J., Cawthray, G. R., Turner, S., Grigg, A. M. and Lambers, H. 2003. Chickpea and white lupin rhizosphere carboxylates vary with soil properties and enhance phosphorus uptake. *Plant and Soil* 248(1/2), 187–197.

Vitousek, P. M., Naylor, R., Crews, T., David, M. B., Drinkwater, L. E., Holland, E., Johnes, P. J., Katzenberger, J., Martinelli, L. A., Matson, P. A., Nziguheba, G., Ojima, D., Palm, C.

A., Robertson, G. P., Sanchez, P. A., Townsend, A. R. and Zhang, F. S. 2009. Nutrient imbalances in agricultural development. *Science* 324(5934), 1519-1520.

Waddell, H., Simpson, R., Henderson, B., Ryan, M., Lambers, H., Garden, D. and Richardson, A. 2015. Differential growth response of *Rytidosperma* species (wallaby grass) to phosphorus application and its implications for grassland management. *Grass and Forage Science* 17, 245-258.

Wang, C., White, P. J. and Li, C. 2016. Colonization and community structure of arbuscular mycorrhizal fungi in maize roots at different depths in the soil profile respond differently to phosphorus inputs on a long-term experimental site. *Mycorrhiza* 27(4), 369-381.

Wang, D., Marschner, P., Solaiman, Z. and Rengel, Z. 2007. Growth, P uptake and rhizosphere properties of intercropped wheat and chickpea in soil amended with iron phosphate or phytate. *Soil Biology and Biochemistry* 39(1), 249-256.

Wang, M., Schafer, M., Li, D., Halitschke, R., Dong, C., Mcgale, E., Paetz, C., Song, Y., Li, S., Dong, J., Heiling, S., Groten, K., Franken, P., Bitterlich, M., Harrison, M. J., Paszkowski, U. and Baldwin, I. T. 2018a. Blumenols as shoot markers of root symbiosis with arbuscular mycorrhizal fungi. *eLife* 7, e37093.

Wang, W., Ding, G. D., White, P. J., Wang, X. H., Jin, K. M., Xu, F. S. and Shi, L. 2019. Mapping and cloning of quantitative trait loci for phosphorus efficiency in crops: opportunities and challenges. *Plant and Soil* 439(1-2), 91-112.

Wang, X., Chen, Y., Thomas, C. L., Ding, G., Xu, P., Shi, D., Grandke, F., Jin, K., Cai, H., Xu, F., Yi, B., Broadley, M. R. and Shi, L. 2017a. Genetic variants associated with the root system architecture of oilseed rape (*Brassica napus* L.) under contrasting phosphate supply. *DNA Research* 24(4), 407-417.

Wang, X., Ding, W. and Lambers, H. 2018b. Nodulation promotes cluster-root formation in Lupinus albus under low phosphorus conditions. *Plant and Soil* 439(1-2), 233-242.

Wang, Y., Krogstad, T., Clarke, N., Øgaard, A. F. and Clarke, J. L. 2017b. Impact of phosphorus on rhizosphere organic anions of wheat at different growth stages under field conditions. *AoB Plants* 9(2), plx008.

Wang, Y. and Lambers, H. 2019. Root-released organic anions in response to low phosphorus availability: recent progress, challenges and future perspectives. *Plant and Soil* 447(1-2), 135-156, https://doi.org/10.1007/s11104-019-03972-8.

Wasaki, J., Maruyama, H., Tanaka, M., Yamamura, T., Dateki, H., Shinano, T., Ito, S. and Osaki, M. 2009. Overexpression of the *LASAP2* gene for secretory acid phosphatase in white lupin improves the phosphorus uptake and growth of tobacco plants. *Soil Science and Plant Nutrition* 55(1), 107-113.

Watt, M. and Evans, J. R. 1999. Linking development and determinacy with organic acid efflux from proteoid roots of white lupin grown with low phosphorus and ambient or elevated atmospheric CO_2 concentration. *Plant Physiology* 120(3), 705-716.

Weaver, D. M. and Wong, M. T. F. 2011. Scope to improve phosphorus (P) management and balance efficiency of crop and pasture soils with contrasting P status and buffering indices. *Plant and Soil* 349(1-2), 37-54.

Weeks, J. J. and Hettiarachchi, G. M. 2019. A review of the latest in phosphorus fertilizer technology: possibilities and pragmatism. *Journal of Environmental Quality* 48(5), 1300-1313.

Weiner, J., Andersen, S. B., Wille, W. K. M., Griepentrog, H. W. and Olsen, J. M. 2010. Evolutionary agroecology: the potential for cooperative, high density, weed-suppressing cereals. *Evolutionary Applications* 3(5-6), 473-479.

Wen, Z., Li, H., Shen, Q., Tang, X., Xiong, C., Li, H., Pang, J., Ryan, M. H., Lambers, H. and Shen, J. 2019. Tradeoffs among root morphology, exudation and mycorrhizal symbioses for phosphorus-acquisition strategies of 16 crop species. *New Phytologist* 223(2), 882–895.

Wen, Z. H., Li, H. G., Shen, J. B. and Rengel, Z. 2017. Maize responds to low shoot P concentration by altering root morphology rather than increasing root exudation. *Plant and Soil* 416(1–2), 377–389.

Withers, P. J., Sylvester-Bradley, R., Jones, D. L., Healey, J. R. and Talboys, P. J. 2014. Feed the crop not the soil: rethinking phosphorus management in the food chain. *Environmental Science and Technology* 48(12), 6523–6530.

Wouterlood, M., Cawthray, G. R., Scanlon, T. T., Lambers, H. and Veneklaas, E. J. 2004. Carboxylate concentrations in the rhizosphere of lateral roots of chickpea (*Cicer arietinum*) increase during plant development, but are not correlated with phosphorus status of soil or plants. *New Phytologist* 162(3), 745–753.

Xiao, K., Katagi, H., Harrison, M. and Wang, Z.-Y. 2006. Improved phosphorus acquisition and biomass production in *Arabidopsis* by transgenic expression of a purple acid phosphatase gene from *M. truncatula*. *Plant Science* 170(2), 191–202.

Xu, C. L., Huang, S. B., Tian, B. J., Ren, J. H., Meng, Q. F. and Wang, P. 2017. Manipulating planting density and nitrogen fertilizer application to improve yield and reduce environmental impact in Chinese maize production. *Frontiers in Plant Science* 8, 1234.

Yan, F., Schubert, S. and Mengel, K. 1992. Effect of low root medium pH on net proton release, root respiration, and root growth of corn (*Zea mays* L.) and broad bean (*Vicia faba* L.). *Plant Physiology* 99(2), 415–421.

Yang, H., Zhang, Q., Dai, Y., Liu, Q., Tang, J., Bian, X. and Chen, X. 2015a. Effects of arbuscular mycorrhizal fungi on plant growth depend on root system: a meta-analysis. *Plant and Soil* 389(1–2), 361–374.

Yang, L., Xie, J., Jiang, D., Fu, Y., Li, G. and Lin, F. 2008. Antifungal substances produced by *Penicillium oxalicum* strain PY-1—potential antibiotics against plant pathogenic fungi. *World Journal of Microbiology and Biotechnology* 24(7), 909–915.

Yang, Z., Culvenor, R. A., Haling, R. E., Stefanski, A., Ryan, M. H., Sandral, G. A., Kidd, D. R., Lambers, H. and Simpson, R. J. 2015b. Variation in root traits associated with nutrient foraging among temperate pasture legumes and grasses. *Grass and Forage Science* 72(1), 93–103.

York, L. M., Carminati, A., Mooney, S. J., Ritz, K. and Bennett, M. J. 2016. The holistic rhizosphere: integrating zones, processes, and semantics in the soil influenced by roots. *Journal of Experimental Botany* 67(12), 3629–3643.

York, L. M., Galindo-Castaneda, T., Schussler, J. R. and Lynch, J. P. 2015. Evolution of US maize (*Zea mays* L.) root architectural and anatomical phenes over the past 100 years corresponds to increased tolerance of nitrogen stress. *Journal of Experimental Botany* 66(8), 2347–2358.

Yu, R. P., Li, X. X., Xiao, Z. H., Lambers, H. and Li, L. 2020. Phosphorus facilitation and covariation of root traits in steppe species. *New Phytology* 226, 1285–1298.

Zak, D. R., Holmes, W. E., White, D. C., Peacock, A. D. and Tilman, D. 2003. Plant diversity, soil microbial communities, and ecosystem function: are there any links? *Ecology* 84(8), 2042–2050.

Zangaro, W., Nishidate, F. R., Vandresen, J., Andrade, G. and Nogueira, M. A. 2007. Root mycorrhizal colonization and plant responsiveness are related to root plasticity, soil fertility and successional status of native woody species in southern Brazil. *Journal of Tropical Ecology* 23(1), 53–62.

Zhang, D., Song, H., Cheng, H., Hao, D., Wang, H., Kan, G., Jin, H. and Yu, D. 2014a. The acid phosphatase-encoding gene *GmACP1* contributes to soybean tolerance to low-phosphorus stress. *PLoS Genetics* 10(1), e1004061.

Zhang, D., Wang, Y., Tang, X., Zhang, A., Li, H. and Rengel, Z. 2019a. Early priority effects of occupying a nutrient patch do not influence final maize growth in intensive cropping systems. *Plant and Soil* 442(1–2), 285–298.

Zhang, D., Zhang, C., Tang, X., Li, H., Zhang, F., Rengel, Z., Whalley, W. R., Davies, W. J. and Shen, J. 2016. Increased soil phosphorus availability induced by faba bean root exudation stimulates root growth and phosphorus uptake in neighbouring maize. *New Phytologist* 209(2), 823–831.

Zhang, L., Fan, J., Ding, X., He, X., Zhang, F. and Feng, G. 2014b. Hyphosphere interactions between an arbuscular mycorrhizal fungus and a phosphate solubilizing bacterium promote phytate mineralization in soil. *Soil Biology and Biochemistry* 74, 177–183.

Zhang, L., Fan, J., Feng, G. and Declerck, S. 2019b. The arbuscular mycorrhizal fungus *Rhizophagus irregularis* MUCL 43194 induces the gene expression of citrate synthase in the tricarboxylic acid cycle of the phosphate-solubilizing bacterium *Rahnella aquatilis* HX2. *Mycorrhiza* 29(1), 69–75.

Zhang, W., Tang, X., Feng, X., Wang, E., Li, H., Shen, J. and Zhang, F. 2019c. Management strategies to optimize soil phosphorus utilization and alleviate environmental risk in China. *Journal of Environmental Quality* 48(5), 1167–1175.

Zhang, X. C., Dippold, M. A., Kuzyakov, Y. and Razavi, B. S. 2019d. Spatial pattern of enzyme activities depends on root exudate composition. *Soil Biology and Biochemistry* 133, 83–93.

Zhu, J. M. and Lynch, J. P. 2004. The contribution of lateral rooting to phosphorus acquisition efficiency in maize (*Zea mays*) seedlings. *Functional Plant Biology* 31(10), 949–958.

Zhu, J. M., Zhang, C. C. and Lynch, J. P. 2010. The utility of phenotypic plasticity of root hair length for phosphorus acquisition. *Functional Plant Biology* 37(4), 313–322.

Zhu, Y. H., Weiner, J., Yu, M. X. and Li, F. M. 2019. Evolutionary agroecology: trends in root architecture during wheat breeding. *Evolutionary Applications* 12(4), 733–743.

Chapter 2

Enhancing phosphorus-use efficiency in crop production

J. L. Havlin, North Carolina State University, USA; and A. J. Schlegel, Kansas State University, USA

1 Introduction

Phosphorus (P) is the second most frequently deficient nutrient, where an estimated 65% (~5.7 billion ha) of global cropland exhibits limited P availability to crops (Batjes, 1997; Hinsinger, 2001). As world population increases to 9.2 billion by 2050, grain production will need to increase by 50% (CAST, 2013). Non-renewable rock phosphate (RP) [$Ca_{10}(PO_4)_6F_2$] is the primary source of fertilizer P, with an estimated 16 billion t of recoverable RP reserves (Schroder et al., 2010). Assuming current consumption rates of 160–180 million t year^{-1}, RP reserves will last ~100 years, with peak RP supply estimated to occur in 2035 (Cordell et al., 2009). Although considerable variation exists in estimates of the remaining RP reserves (Table 1), increasing crop recovery of applied P and reducing soluble and sediment P transport from cropland are increasingly important to conservation of RP reserves. For example, Li et al. (2018) estimated RP depletion within 70-140 years; however, with effective P management to enhance P recovery and limit P losses, depletion time can be increased by 50 years. Van Kauwenbergh (2010) reported revised USGS estimates of ~70 billion t RP reserves and estimated ~300 years of recoverable RP (Table 1). In addition, the Western Phosphate Field located in the northern

http://dx.doi.org/10.19103/AS.2019.0062.08

Table 1 Estimated years remaining of global rock phosphate reserves

Years remaining	References
70-140	Li et al. (2018), Fixen and Johnston (2009), Smit et al. (2009)
100	Heffer et al. (2006), Smil (2000)
130	Cordell et al. (2009)
200	Herring and Fantel (1993)
130-200	Van Vuuren et al. (2010)
343	Roberts and Stewart (2002)
300-400	van Kauwenbergh (2010), IFDC (2010)
600-1000	Scholz et al. (2015), Isherwood (2003)

Rocky Mountains contains an estimated 825 billion t and is the largest RP deposit on Earth. Although costly to extract at ~1800 m depth, the estimated annual production of 200 million t would meet P demand for ~1000 years (Scholz et al., 2014). Regardless of the estimate, RP depletion is not imminent; however, with increasing demand for P fertilizer (with greater food demand), the importance of increasing soil P availability while reducing soil P depletion by off-site P transport is crucial (Cordell and White, 2013; Roberts and Johnston, 2015).

Conserving RP resources requires enhanced P efficiency in production agriculture and throughout the RP processing system (Scholz et al., 2015). In the year fertilizer P is applied, crop P recovery is commonly considered to be <25% and the availability of the remaining >75% fertilizer P in the soil to subsequent crops is relatively limited. Understanding fertilizer P retention or fixation reactions in soil is crucial to developing and adopting P management practices that increase phosphorus-use efficiency (PUE). The following sections provide a review of soil, plant, and management factors important to increase crop recovery of applied P.

2 Phosphorus-use efficiency

Crop recovery of applied nutrients is routinely quantified to assist researchers in evaluating soil and crop management practices that can maximize crop nutrient use and minimize residual nutrients as potential environmental contaminants (Sims and Kleinman, 2005). A number of methods have been employed to represent PUE (Syers et al., 2008). The *direct method* measures crop recovery of ^{32}P- or ^{33}P-labeled fertilizer and is the most accurate method for quantifying PUE of freshly applied P. Unfortunately, the half-lives of ^{32}P and ^{33}P are too short (14 and 25 days, respectively) for use in long-term P response studies to estimate residual available P. Crop P recovery with the *direct method* commonly ranges from 5% to 25%, where the remaining 75-95% of crop P is absorbed from native soil P or residual fertilizer P sources.

The most common method used to quantify recovery of applied P is the *difference method*. When based on crop P uptake, *apparent recovery* (AR) is defined by:

$$AR = \frac{\text{crop P uptake (fertilized soil)} - \text{crop P uptake (unfertilized soil)}}{\text{P applied}}$$

When the *difference method* is based on crop yield, the *agronomic efficiency* (AE) is defined by:

$$AE = \frac{\text{crop yield (fertilized soil)} - \text{crop yield (unfertilized soil)}}{\text{P applied}}$$

The *difference method* can also be used to quantify the *physiological efficiency* (PE) defined by:

$$PE = \frac{\text{crop yield (fertilized soil)} - \text{crop yield (unfertilized soil)}}{\text{crop P uptake (fertilized soil)} - \text{crop P uptake (unfertilized soil)}}$$

The recovery of applied P determined by the *difference method* requires inclusion of an unfertilized treatment; thus, **AR** or **AE** values are commonly low when native soil P sources are high. Under low soil P availability, crop recovery of applied P is greater. For these reasons, the *difference method* is best used for crop N recovery because residual fertilizer N is usually much lower than for P.

Alternatively, a *balance method* has been more recently used to quantify *utilization efficiency* (**UE**) and is defined by:

$$UE = \frac{\text{crop P uptake (fertilized soil)}}{\text{P applied}}$$

Compared to **AR**, the **UE** reflects crop recovery of both added P and residual soil P reserves, which more accurately reflects P availability. **AE** represents the additional yield obtained per unit of applied P and, therefore, reflects the relative responsiveness of the site to P fertilization. **PE** represents the additional yield obtained per unit of additional P uptake (yield and P uptake due to fertilizer only) and indicates how efficiently a plant utilizes applied P in accumulating biomass and/or grain. Neither **AE** nor **PE** nor **UE** is routinely reported (Ciampitti and Vyn, 2014).

As expected, the wide variation in reported values for P efficiency is a simple artifact of different methods of expression as well as variation between sites and years within the same site. To illustrate differences between P methods of quantifying PUE, data from Hossain and Sattar (2014) and Schlegel and Havlin

(2017) were used to calculate **AR**, **UE**, **AE**, and **PE** as defined above (Table 2). At all four sites, **AR** < **UE** because residual soil P uptake is not included in **AR**, whereas with **UE** crop, P uptake includes fertilizer P + residual soil P. Syers et al. (2008) documented that the *balance method* was preferred over the *difference method* because the latter did not reflect the residual value of previous P application on crop P and yield. The **AR** was significantly higher with irrigated corn and sorghum (sites C–D) than at sites A and B (rainfed wheat) and is due to higher P uptake with irrigated crops combined with 30% lower P rates. In addition, **UE** at sites C and D were > 1.0, which means crop P uptake was greater than P rate applied, compared to sites A and B (Table 2). Similar results were reported from estimated global grain yield and P use (Fig. 1). These data illustrate **AR** values are commonly < 0.25, while **AR** > 0.5 indicates crop P removal > P applied and should result in declining soil P availability (Dhillon et al., 2017). Generally, when **UE** > 1.0, soil P reserves and plant-available P are being depleted, whereas **UE** < 1.0 represents a buildup of soil P reserves (Syers et al., 2008; Roberts and Johnston, 2015). Long-term P management where **UE** < 1.0 potentially increases risk of P transport to surface and ground waters (Heathwaite et al., 2005).

Table 2 Examples of common parameters used to quantify phosphorus-use efficiency[a]

Parameter	Location[b]			
	A	B	C	D
Bray P (mg kg^{-1})	14	31	7	9
Bray P interpretation	Medium	High	Low	Low
Fertilizer P rate (kg ha^{-1})	30	30	20	20
Grain yield (t ha^{-1})				
Unfertilized	1.5	1.8	5.9	6.7
Fertilized	2.0	2.4	10.4	8.2
P uptake (kg ha^{-1})				
Unfertilized	3.9	8.9	30.5	15.8
Fertilized	5.9	12.7	41.9	25.6
Apparent recovery (AR) (kg kg^{-1}) $[P_{uptake}(+P) - P_{uptake}(-P)]/P_{applied}$	0.07	0.13	0.57	0.49
Utilization efficiency (UE) (kg kg^{-1}) $[P_{uptake}(+P)]/P_{applied}$	0.20	0.42	2.09	1.28
Agronomic efficiency (AE) (kg kg^{-1}) $[\text{yield}(+P) - \text{yield}(-P)]/P_{applied}$	15.6	20.7	221	76
Physiological efficiency (PE) (kg kg^{-1}) $[\text{yield}(+P) - \text{yield}(-P)]/[P_{uptake}(+P)-P_{uptake}(-P)]$	230	160	390	160

[a] See text for complete definition of AR, UE, AE, PE.
[b] A,B = rainfed wheat (Hossain and Sattar, 2014); C = irrigated corn (Schlegel and Havlin, 2017); D = irrigated sorghum (Schlegel and Havlin, 2019, *pers. comm.*)

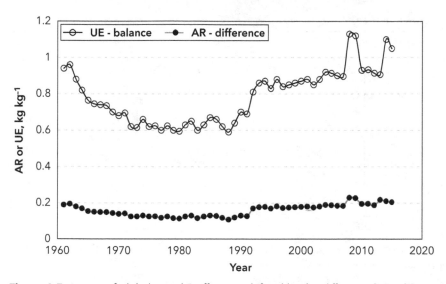

Figure 1 Estimates of global cereal P efficiency defined by the *difference* [AR = [(P$_{uptake}$ (+P) − P$_{uptake}$ (−P)]/P$_{applied}$] and *balance* [UE = P$_{uptake}$(+P)/P$_{applied}$] methods. Source: adapted from Dhillon et al. (2017).

3 Soil factors

3.1 Buffering rhizosphere/solution phosphorus

The dynamics of P supply to plant roots are dominated by processes between the soil P buffer capacity (BC) and the rhizosphere. Plant roots absorb $H_2PO_4^-$/ HPO_4^{-2} in soil solution, which is replenished or buffered by P pools of variable quantities and kinetics of desorption or dissolution, collectively termed P BC (Fig. 2). Soil solution P concentration ranges from 10^{-4} to 10^{-6} M (3.1-0.031 mg P L^{-1}), which represents ~0.2 kg P ha-30 cm^{-1} soil (assumes 20% Θ_v, 0.31 mg P L^{-1}, 0.3 m depth). With a typical crop P requirement of 20 kg P ha^{-1}, solution P must be replenished 100 times during the growing season. Plant P uptake primarily occurs through plant root hairs within the rhizosphere soil, which occupies a soil volume ~1 mm from the root surface (Fig. 2). The extent of root hair development and rhizosphere volume varies greatly with plant species, soil chemical and physical properties, and solution P concentration, where increasing solution P decreases root hair length (Fig. 3). Given that mass flow and root interception contribute <5% of P supply to plant roots, P diffusion dominates P transport from the bulk soil to and within the rhizosphere (Kovar and Claassen, 2005).

With ~0.005–0.015% of total P present in solution, meeting plant P demand requires desorption or dissolution of mineral P. As added $H_2PO_4^-$/HPO_4^{-2} is increasingly adsorbed to Al/Fe oxide and other mineral surfaces, precipitated

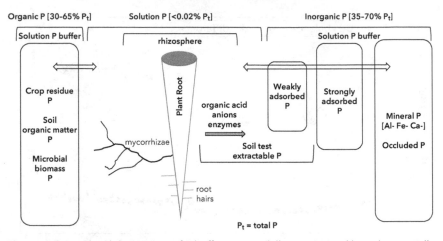

Figure 2 Generalized description of P buffering in soil illustrating readily and potentially available inorganic and organic P reserves.

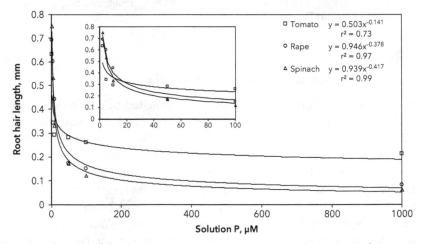

Figure 3 Influence of solution P concentration on average root hair length of three plant species. Figure inset shows 0–100 µM solution P range. Source: adapted from Fohse and Jungk (1983).

as Al/Fe-P and Ca-P minerals, and adsorbed to organic matter (OM) surfaces or immobilized by microbes, access to soluble P by plants roots is reduced (Fig. 2). Soluble fertilizer or manure P applied to soils is readily converted to less soluble forms (adsorbed and/or precipitated) with decreased availability to plants, which reduces recovery of applied P. Globally, annual P application represents 1-2.5% total soil P compared to 0.2-0.6% of total P removed by crop P annually (Hesterberg, 2010).

Depending on soil OM and clay content, total soil P comprises 30-60% organic P and 35-70% inorganic P. Inorganic P represents P adsorbed to mineral surfaces with variable strengths of adsorption and mineral P of variable solubility (Fig. 2). In acid soils, Al/Fe P minerals dominate, and P solubility increases with increasing pH, whereas in neutral and calcareous soils Ca-P minerals dominate, and P solubility generally decreases with increasing pH (Lindsay, 1979). Recent advances in spectroscopic analyses of soil P fractions show that Ca-P minerals are found in acid soils in contrast to predictions from P solubility principles (Beauchemin et al., 2003).

With plant P uptake rates ranging from 0.1 to 0.6 kg P ha^{-1} d^{-1}, solution P concentration in the rhizosphere is rapidly decreased and must be frequently replenished from the P BC to meet plant P demand (Leigh and Johnston, 1986; Syers et al., 2008). Hendriks et al. (1981) demonstrated the P depletion profile, where solution and exchangeable P decreased 3- and 2-fold in the rhizosphere, respectively, compared to the bulk soil. Ultimately, P deficiency in plants occurs when the P BC is insufficient to replenish solution P concentration in the rhizosphere. While adequate P buffering kinetics is important throughout the growing season, it is especially important early in plant growth (Mengel and Barber, 1974; Grant et al., 2001).

Soil P BC comprises surface exchange, mineral solubility, and mineralization of OM fractions in soils and is defined as (Kovar and Claassen, 2005; Havlin et al., 2014):

$$BC = \frac{dC_M}{dC_S}$$

where, dC_M is the solution P concentration + soil matrix P (adsorbed, mineral, organic) and dC_S is solution P. Figure 4 illustrates BC for four soils with variable chemical and physical properties (Kovar and Barber, 1988). In this example, BC decreases in the order A > B > C > D.

Soils with higher BC generally have greater clay, Al/Fe oxide, CaCO$_3$, and/or OM contents. As P is added, solution P increases more in soils with a lower BC compared to high BC soils (Fig. 4). Alternatively, as solution P is removed by plant uptake, soils with higher BC have greater potential to resupply solution P and meet P uptake demand. As dC_S increases, dC_M decreases, therefore, for a specific soil, BC is less at higher solution P concentrations.

With fertilizer or manure P applications, soluble P is converted to less soluble P forms (C_M) defined by their relative capacity to resupply solution P (Fig. 5). Application of P fertilizer in excess of crop P demand increases P saturation of the soil BC. Soluble P from fertilizer or manure/waste P applications

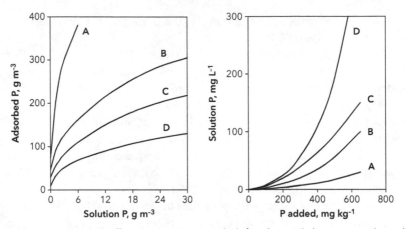

Figure 4 Generalized P buffer capacity (BC) in soils (left), where BC decreases in the order A > B > C > D. When soluble P is added to soil (right), solution P increases with decreasing BC in the order D > C > B > A. Source: adapted from Kovar and Barber (1988).

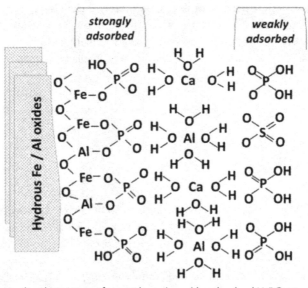

Figure 5 Generalized structure of strongly and weakly adsorbed $H_2PO_4^-$ to Al–O or Fe–O surfaces. Source: adapted from Hedley and McLaughlin (2005).

adsorbs to soil mineral and OM surfaces, displacing other anions with lower affinity. Adsorption of P on Al/Fe oxides and $CaCO_3$ takes hours to days to occur, whereas precipitation as $Al/FePO_4 \cdot 2H_2O$ and/or $CaHPO_4 \cdot 2H_2O$ takes months (Hedley and McLaughlin, 2005). These P reserves are potentially plant available, although their relative contribution to plant-available P is difficult to quantify. Enhancing P soil testing methods to account for the contribution of mineral P dissolution to plant-available P may reduce over-application of P and the demand for fertilizer P.

3.2 Quantifying residual soil phosphorus

Chemical extraction methods are generally used to quantify each component of residual P (Table 3; Hedley et al., 1982). For example, Couto et al. (2017) applied ~500 kg P ha^{-1} as manure and fertilizer over 12 years and reported significant increases in soluble, readily available, and slowly available P reserves compared to unfertilized soil. In addition, organic P fractions were increased with manure application compared to inorganic P fertilizer. Even though P fractionation methods are widely used to distinguish P pools by their relative supply of plant-available P (Condron and Newman, 2011), the contribution of individual fractions to plant P uptake is difficult to quantify. Common soil test methods used to extract soluble and readily available P fractions are well established and are correlated with crop response to applied P, crop P removal, and provide the basis for fertilizer and manure P recommendations. Therefore, the primary driver to replenish solution P from labile and non-labile P is the soil's P status as measured by accepted soil test extraction, which should represent relative P availability to plants.

Accurately estimating the quantity of plant-available P in soil requires quantifying plant-available P from both residual and applied P. Residual soil P fractions accumulate with P applications in excess of crop P removal, which represents the difference between P inputs (fertilizer, manure, weathering, deposition) and P outputs (crop P removal, runoff/erosion). Improving our ability to quantify the contribution of residual P to plant-available P through

Table 3 Selected soil extractants used to quantify inorganic P (P$_i$) and organic P (P$_o$) components in the soil pool of available P

	Relative P availability	Extractants	P component or form
Decreasing extractability	High-soluble P	H_2O, resin	Solution P ($H_2PO_4^-$, HPO_4^{-2}, P$_o$)
	Readily available	0.5 M NaHCO$_3$	Weakly adsorbed P$_i$ readily oxidizable P$_o$
	Moderately available	0.1 M NaOH	Moderately adsorbed to Al/Fe
		1 M HCl	Moderately adsorbed to Ca
	Non-available	0.5 M NaOH	Strongly adsorbed P$_i$ weakly oxidizable P$_o$
	Residual	H_2SO_4, H_2O_2, HCl	Residual P$_i$ + P$_o$

Source: adapted from Hedley et al. (1982).

P desorption from mineral surface to solution in needed (Menezes-Blackburn et al., 2016).

For example, following an initial application of 145 and 290 kg P ha^{-1}, soil test level increased 2-fold and 4-fold, respectively, above the unfertilized soil (Fig. 6; Webb et al., 1992). Over 14 years of corn–soybean production, soil test levels declined below the critical level (20 mg Bray-1 P kg^{-1}) after 8 and 12 years with the 145 and 290 kg P ha^{-1} rates. More importantly, economic responses to P were only observed when soil test levels were below the critical level (data not shown). Similar results were reported by Karpinets et al. (2004). Using residual P data from a long-term study at Rothamsted, Johnston et al. (2014) quantified the decline in NaHCO$_3$-extractable P over 16 years and demonstrated a 50% decrease in soil test P in 9 years of crop P removal and estimated that soil test P would decrease below the critical level in 25 years. With fertilizer or manure P applied at rates that increase soil test P above the critical level, the crop yield response to applied P was low (Medinski et al., 2018). Figure 7 illustrates the effects of soil test P on relative crop response to applied P.

In a 39-year study, Selles et al. (2011) reported NaHCO$_3$-extracted P increased 0.15 kg ha^{-1} per kg P ha^{-1} applied in excess of crop P removal. When crop P uptake < P applied, soil test P increased; in contrast, it decreased when crop P uptake > P applied (Fig. 8). With fertilizer P withheld during the last 12 years of the study, crop P uptake was 90–105% of previous P applications, demonstrating long-term plant availability of residual labile and non-labile P reserves. Similarly, Syers et al. (2008) reported Mehlich-1 P increased 1 mg kg^{-1} with each 10 kg P ha^{-1} applied. Alternatively, common soil tests (e.g. Mehlich-3, Bray-1 P) used to predict probability of crop response to annual P applications

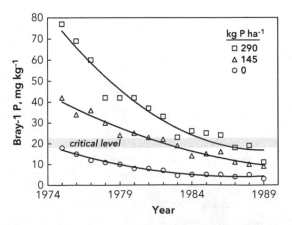

Figure 6 Bray-1 P levels following one-time P application (in 1974) to corn–soybean rotation. Shaded band represents critical soil test level = 20 mg Bray-1 P kg^{-1} soil. Source: adapted from Webb et al. (1992).

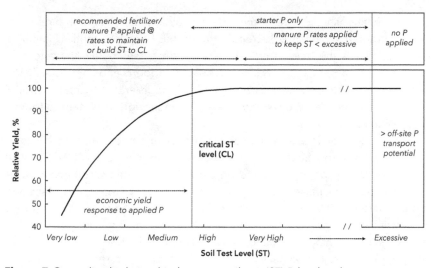

Figure 7 Generalized relationship between soil test (ST) P level and crop response to applied P. When soil test P < CL (critical soil test P level), P should be applied at rates to achieve ~95% yield; when soil test P = CL, P should be applied at rates to maintain CL; commonly there is no crop response to applied P when soil test P > CL, and there is an increased potential for off-site P transport.

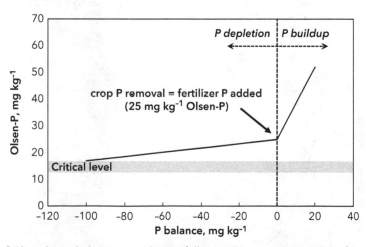

Figure 8 Phosphorus balance over 12 years following 27 years (1967–1994) of P applied annually at 10 kg P ha^{-1} to winter wheat (no P applied after 1994). Soil test declined when grain P removal > P added. Source: adapted from Selles et al. (2011).

are equally effective as P fractionation methods in assessing impacts of long-term P applications on crop P uptake (Motavalli and Miles, 2002). In this study it is interesting to note that after 11 years of P application, manure P increased total and organic P fractions more than fertilizer P additions.

The relationship between the quantity of applied P and the resultant change in soil test P relative to crop P removal has been frequently quantified. For example, as the ratio of crop P removal decreases below the rate of P applied, soil test P substantially increases (Fig. 9; Roberts and Johnston, 2015). Therefore, as soil test P increases, reduction in P inputs and/or increased crop P removal (elevated yields, cover crops, etc.) is increasingly important to reduce off-site P transport.

As a result of intensive animal production methods, manure disposal is relegated to crop fields near the source (Kellogg et al., 2000). Because manure rates are commonly based on crop N requirements, soil test P levels can substantially increase, resulting in enhanced P loss potential to surface and ground waters (USDA-NRCS, 2011). As shown in Fig. 7, as soil test P increases to *very high* or *excessive* levels, manure application rates should be limited to the estimated quantity of crop P removed from the field (Kleinman et al., 2017). Long-term P fertilization, especially with high P rates common to intensive animal production, substantially increases P saturation, which reduces P retention capacity. Phosphorus loss assessment (P index) tools have been developed to address elevated residual P transport problems in manure-amended soils, which include many other factors besides soil testing (Sharpley et al., 2012). For example, a well-established management practice is using riparian buffers to reduce sediment-bound P loss to surface waters (Vidon et al., 2018).

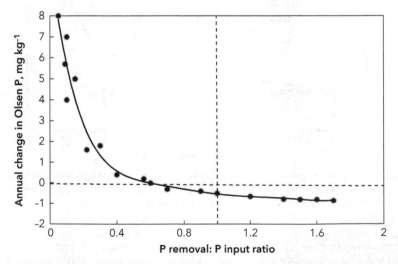

Figure 9 Effect of crop P removal to P input ratio on annual change in Olsen P soil test level from long-term studies in the UK and the US. Source: adapted from Roberts and Johnston (2015).

4 Plant factors

4.1 Opportunities in plant genetics

Given the scenario of decreasing fertilizer P resources, improving P efficiency includes advancing the capacity of plants to acquire soil P through improved plant genetics primarily related to plant root architecture and P metabolism and relevant physiological processes (Niu et al., 2013; van der Wiel et al., 2016). Enhancing internal P efficiency through genetic manipulation to improve the efficacy of P allocation and mobilization within the plant is an essential long-term goal in plant genetics (Parentoni and Junior, 2008; Wang et al., 2010a; Rose and Wissuwa, 2012).

In many plant species and cultivars, root-to-shoot ratio increases under limited P availability, which increases PUE (Wang et al., 2010b; Neto et al., 2016). As P supply in the rhizosphere decreases during early plant growth, lateral and axial root growth may increase, accompanied by decreased primary root growth in some species (Bovill et al., 2013). Increased root hair growth may also occur in some plants, which would conserve plant C compared to lateral root expansion (Wang et al., 2004; Lynch, 2007; Richardson et al., 2011). Altering root architecture by increasing root density and reducing axial root angle would increase root length and subsequent access to higher soil test P levels in the surface soil (Ao et al., 2010). Although not all crops have the capacity to increase root hair density under low P supply, Yan et al. (2004) and Wang et al. (2004) reported increased root hair length and density in common bean and soybean. Plant breeding to alter root architectural traits for increased PUE in the surface soil may come at the expense of deeper rooting characteristics important for accessing subsoil water.

Genetic variation between and within plant species in root exudation of organic acids has been demonstrated to enhance solubility and mobilization of soil P (Vance et al., 2003; Yan et al., 2004; Badri and Vivanco, 2009). There are several probable mechanisms likely to enhance P solubilization by rhizosphere microbes, including organic acid anion release and root exudation of phosphatase enzymes (Marschner, 1998). For example, Krishnapriya and Pandey (2016) evaluated 116 soybean genotypes and found a wide range in P acquisition efficiency, whereby under low P, the efficient genotypes exuded greater quantity of organic acid anions in the rhizosphere, increasing P acquisition. Organic acid anion exudation also increases rhizosphere microbial growth, contributing to P solubilization (Wang et al., 2010a). For example in soybean, oxalate and malate exudation was increased under low soil P, whereas citrate exudation occurred under low pH, reducing Al toxicity (Wang et al., 2010b). Organic acid chelation of soluble Al^{3+} to reduce Al toxicity may enhance P uptake by improved root growth and function (Delhaize, 2001; Mariano et al.,

2005; Shabnam and Iqbal, 2016). Increasing acidity in the rhizosphere also functions to increase solution P in calcareous soils (Korkmaz et al., 2009).

Inositol phosphate (phytate) comprises the majority of organic P in soils and is relatively stable (Singh and Satyanarayana, 2011). Although phosphohydrolase enzymes are common, phytate mineralization to plant-available P in the rhizosphere is limited by relatively low phosphatase and phytase activity (Greiner, 2007; Sun et al., 2017). Therefore, enhancing microbial exudation and/or extracellular root excretion of phytase in the rhizosphere may increase phytate mineralization and P uptake (Richardson et al., 2009). Phytase enzyme exudation by plant roots is genetically controlled; therefore, development of transgenic plants that increase phytase concentration in the rhizosphere through direct enzyme release by roots or release of organic acid anions that increase microbial exudation of phytase should be considered (Richardson et al., 2011; Balaban et al., 2017).

Most plants have evolved with the capacity to form root–fungal symbioses (Fig. 2). Mycorrhizal associations substantially increase the volume of soil explored by root–mycorrhizae associations, increasing access to water and nutrients. Plants growing on soils nearly depleted of plant-available P can complete their life cycle through root mycorrhizal associations (Marschner, 1998). Koide and Kabir (2000) provided evidence that fungal hyphae may access soil P reserves not accessible to plant roots. Even though P acquisition by mycorrhizae decreases with increasing soil P, the extent and effectiveness of mycorrhizal colonization vary widely among plant species (An et al., 2010). Therefore, exploiting genetic control of mycorrhizal infections and efficacy in accessing soil P can enhance P acquisition from soil.

Enhancing internal P efficiency requires increasing P transport kinetics from mature leaves to newly developing leaves. The primary focus should involve manipulation of relevant P transporter genes. Recent research with rice identified several nucleic acid-interacting proteins responsible for differences in internal P efficiency (Wissuwa et al., 2015). Although difficult to engineer, another potential strategy to enhance internal P efficiency is to adapt P metabolism to a low P requirement. The options might include substitution of membrane phospholipids with sulfo- or galacto-lipids and/or potential for phosphate replacement with phosphite (Simeonova et al., 2010). Details of these potential genetic manipulations have been summarized (Veneklass et al., 2012; van der Wiel et al., 2016). Clearly, plants grown in low-P soils exhibit both external and internal responses to P starvation, which are genetically controlled. Even though many of these genes have been identified, genetic manipulations to improve internal PUE is still difficult. Continued genetic research is crucial to improve internal PUE; however, in the short term it may be more practical to focus on the efficiency of acquiring soil P (Bovill et al., 2013; Mendes et al., 2014; Heuer et al., 2017).

5 Management factors

5.1 Phosphorus rate

Soil test methods for P have been developed to represent an index of plant-available P, where extractable P represents the relative P supply from soil. Phosphorus recommendations based on soil test P are well established and reliable (Havlin et al., 2014). Unfortunately, P recommendation models should be continually re-evaluated to reflect advances in plant genetics and soil/crop management technologies that affect crop P requirements. Clearly, P applied in excess of crop requirement results in increased residual P reserves that, if not utilized or recovered in subsequent crops, may result in off-site P transport contributing to degradation of water quality (Sattari et al., 2012). This is especially important with continual animal or other waste applications based on crop N requirements, where P applications often exceed crop P demand (Sims and Kleinman, 2005; Abit et al., 2018).

Variable rate application technologies can substantially influence P efficiency by reducing the variability in soil test P throughout a field; however, the economic advantage of variable compared to uniform P management depends on the distribution of soil test P and the percentage of the field testing below the critical soil test P level. Within-field spatial variability in soil test P is well documented (Havlin and Heiniger, 2009). Application of P to those areas below the soil test P critical level and withholding P from areas above the critical level ultimately result in reducing the spatial variability in soil test P (Wittry and Mallarino, 2004). This study also documented a 12–41% reduction in total P applied with variable compared to uniform P applications. In the same study, soil test P levels increased where manure was applied to regions below the soil test P critical level, whereas soil test P levels increased in regions where manure was applied to soils testing at or above the critical P level.

Using an on-the-go near infrared/visible sensor to quantify the spatial distribution of soil test P at a 1 m^2 resolution, Malecki et al. (2008) demonstrated significantly higher corn yields and reduced variability in soil test P compared to field-average soil sampling and uniform P application. Although traditional soil sampling at 1 m^2 resolution is not practical, grid-sampling at <1 ha can sufficiently quantify the spatial distribution of soil test P and support variable vs. uniform P application decisions (Schepers et al., 2000; Havlin and Heiniger, 2009).

5.2 Phosphorus sources

In addition to enhancing phytase exudation in the rhizosphere by plant roots or microbes through genetic engineering (Balaban et al., 2017), development of bio-fertilizers containing live microorganisms has been demonstrated to enhance P mineralization and P availability to plants (Fuentes-Ramirez, 2005). Either direct soil application or as a seed inoculant have been documented (Li and Zhang, 2001; Idriss et al., 2002; Mazid and Khan, 2014). For example, Hossain and Sattar (2014) reported similar wheat yield and P uptake with 15 kg P ha^{-1} + seed coated with either *Klebsiella* or *Pseudomonas* sp. (P-solubilizing bacteria), compared to 30 kg P ha^{-1} without the bacteria. Vanneeckhaute et al. (2016) documented that bio-fertilizers represent promising alternatives to inorganic P fertilizers in enhancing PUE. In a recent literature review, Schutz et al. (2017) demonstrated 8–22% increased crop response to bio-based fertilizers, where (1) yield increases in arid climates > tropical climates > oceanic climates > continental climates; (2) arbuscular mycorrhizal fungi > N$_2$ fixing + P-solubilizing bacteria > N$_2$ fixing bacteria > P-solubilizing bacteria; (3) medium soil P > low soil P; and (4) legumes > cereals > root crops.

Numerous studies have documented the advantage of liquid over granular P sources (Lombi et al., 2006; McBeath et al., 2007). McLaughlin et al. (2011) demonstrated under field conditions that fluid fertilizer P increased P efficiency on highly calcareous soils. Crop recovery of applied P is frequently increased with addition of N (Djinadou et al., 1995; Schlegel and Havlin, 2017). These data show that adding urea ammonium nitrate with ammonium polyphosphate substantially increased soil test P (particularly in acid soils compared to a calcareous soil) and is related to enhanced P retention commonly observed in Ca-dominated soils (Fig. 10).

With increasing global population, animal and biosolid wastes provide substantial resources for nutrient recycling in agriculture. For example, struvite

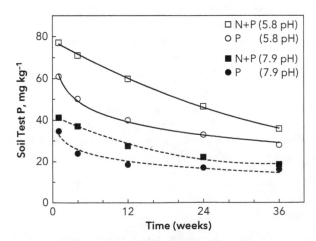

Figure 10 Influence of 40 kg P ha^{-1} applied with or without N on soil test P in a slightly acid (pH 5.8, 10 mg Bray-1 P kg^{-1}) and calcareous (pH 7.9, 7 mg Olsen P kg^{-1}) soils; soil pH measured in 1:1 soil:water. Source: adapted from Djinadou et al. (1995).

($MgNH_4PO_4 \bullet 6H_2O$) has been successfully produced from diverse wastes and provides a valuable slow-release P fertilizer (Rahman et al., 2014; Talboys et al., 2016; Yetilmezsoy et al., 2017). Incinerated animal waste ash has been demonstrated to provide soluble P to field crop production (Crozier et al., 2009). In addition, composted plant residues and animal waste materials mixed with RP have been demonstrated to enhance P availability compared to RP alone on severely acid soils (Oyeyiola and Omueti, 2016).

Another technology claimed to improve P efficiency by reducing P retention reactions is AVAIL®, which is a *maleic-itaconic acid* copolymer impregnated into fertilizer P granules or added to liquid P sources. The ability of AVAIL® to increase crop recovery of added P by reducing P retention reactions in soil has been reported (Hopkins, 2013), although others have reported little or no response (Chien et al., 2014). Review of the literature reveals that many of the studies reporting little or no response were conducted on soils at or near critical soil test P levels or P rates used were too high. Thus, crop responses to lower P rates with AVAIL® were often observed compared to higher P without AVAIL®. Although not completely elucidated, sequestering Al, Fe, and Ca to reduce adsorption and precipitation of added P has commonly described the mode of action. Doydora et al. (2017) recently reported increased soluble P with added AVAIL® that was due to competitive adsorption of COO^- and $H_2PO_4^-$ to Al and Fe oxide surfaces, with slightly greater adsorbed P to Fe oxide. These results also concluded that P solubility effects would be enhanced by minimizing the soil volume in contact with applied P + AVAIL®.

5.3 Phosphorus timing

As with other macronutrients, P should be applied at times synchronous with peak plant growth and nutrient demand (Havlin et al., 2014). With many crops, P uptake demand decreases with crop maturity; therefore, maintaining higher solution P concentrations near developing roots is critical for optimum crop response to applied P (Mengel and Barber, 1974; Schroder et al., 2010. Grant et al. (2001) documented the importance of early-season P nutrition in numerous field studies with grain crops. Obviously, there are examples where P applied later in the season can influence P nutrition, for example, in forages, turfgrass, and other solid-seeded crops where surface root density is high and plant-available water is optimal (e.g. irrigation).

5.4 Phosphorus placement

Numerous studies have been conducted to quantify the fertilized soil volume needed to maximize crop P uptake (Kovar and Barber, 1987, 1989; Su and Miller, 1993). These data showed maximum P uptake occurred with 2-25%

fertilized soil volume, where the required P-fertilized soil volume decreased with increasing soil BC. Although soils exhibit variable P retention capacities, band-applied P substantially increases solution P concentration in the vicinity of the band, which decreases as P diffuses into unfertilized soil. As the volume of P-fertilized soil increases, root surface area in contact with fertilizer increases, which increases P uptake. As the same P rate is mixed with more soil, additional soluble P is adsorbed by soil into poorly available forms (Fig. 2), reducing P uptake (Kovar, 2001). Although total daily P uptake is low in young plants, the P uptake rate per unit of root length is high (Mengel and Barber, 1974). Under these conditions, elevated solution P concentrations should provide sufficient P diffusion to the root surface (Alam et al., 2018). Thus, positioning P close to the expanding root system enhances P uptake and PUE. In pasture and forage crops, early-season broadcast P fertilizer can enhance early P uptake and plant growth compared to banded P; however, similar yields occur with banded P application later in the season (Sweeney et al., 1996).

A summary of 40 years of P placement research demonstrated corn grain yield when broadcast P was ~90% of band-applied P, and grain responses to band P occurred more frequently at lower (<15 kg P ha^{-1}) P rates, provided soil test P levels were below the critical level (Bly et al., 2015). In calcareous soils, Rehim et al. (2012) reported increased wheat yield and P UE with banded P compared to broadcast P (Fig. 11). In a recent review and meta-analysis of over 1000 data sets from nutrient placement studies, Nkebiwe et al. (2016) reported 27% and 15% higher grain yield and N and P uptake with band applied urea + H$_2$PO$_4^-$ and NH$_4^+$ + H$_2$PO$_4^-$ compared to broadcast. Placement responses were significantly greater with N + P treatments than with N or P alone. In addition, deeper placement (~10 cm) resulted in greater placement responses

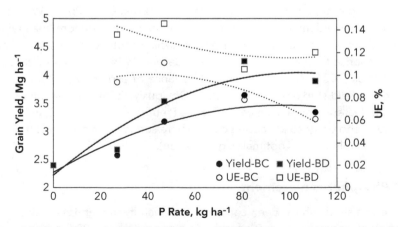

Figure 11 Effect of broadcast (BC) and band-applied (BD) P on grain yield and utilization efficiency (UE) in winter wheat. Source: adapted from Rehim et al. (2012).

compared to surface or shallow band placement and was likely due to greater plant-available water with soil depth.

A thorough review by Schroder et al. (2010) confirms enhanced P efficiency by positioning fertilizer or manure P sources in proximity to the expanding root system. Many studies have suggested broadcast P rates should be double band-applied P to obtain similar crop yield (Sander et al., 1990; McKenzie and Roberts, 1990; van Dijk, 2003; Syers et al., 2008; Barbieri et al., 2014). Peterson et al. (1981) reported greater P response in winter wheat to banded P compared to broadcast P in low soil test P soils, although as soil test P increased to reach critical soil test P levels, crop response to placement was not observed.

Crop responses to P placement can be described by a generalized response to either banding or broadcast P (Fig. 12). In the first scenario (A), crop response to broadcast is slightly better than banding P, which likely occurs with solid-planted crops (forages, turf, etc.) where sufficient roots exist near the soil surface, and band spacing might be too wide to provide sufficient P to meet early crop demand. In fields where soil test P is at or above the critical level, soils are warm and moist, and P is applied to warm-season crops, yield response can be similar between broadcast and band-applied P (scenario B). More commonly, crop yield response to banding P is greater than observed with broadcast P unless sufficiently high broadcast P is applied (scenario C), which likely occurs under low soil test P and in soils with high P retention capacity under cool, wet conditions. Similarly, with early-season crops planted in cool, wet, low-P soils, even high broadcast P rates may not produce similar yield as band-applied P (scenario D). It is likely that under these conditions, sufficiently high broadcast P could be applied to optimize yield; however, P

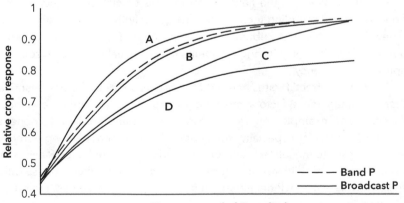

Figure 12 Generalized crop yield responses to P placement, where broadcast > banding (A); broadcast = banding (B); banding > broadcast except at high P rates (C); and banding > broadcast regardless of P rate (D). see text for discussion.

efficiency would be low with elevated residual P reserves. Many other soil and crop management factors strongly influence crop response to P placement. For example, seedling vigor and ultimately yield may be enhanced by band-applied P compared to broadcast P in soils where soil compaction, reduced and no-tillage management, or other management or environmental conditions reduce seedling vigor (Havlin et al., 2014).

6 Conclusion and future trends

Significant progress in enhancing P efficiency in production agriculture requires improved prediction of plant-available P from residual soil P and crop response to applied P in the year of application. Regardless of which parameter is used to quantify P efficiency, P applied at rates greater than estimated crop P removal ultimately increases residual P reserves and their potential transport to surface waters. Understanding P BC and residual P availability greatly improves P efficiency and conserves limited RP resources. This is particularly critical with manure P annually applied at rates above crop requirement, which demands distribution of manure P to P-deficient land areas and/or manure P application rate based on the crop P requirement.

Accurately monitoring soil test P variability within fields is crucial to identifying additional P needed to meet crop P demands. When at or slightly above the critical soil test P level, P rates equivalent to crop P removal should maintain the critical P level. When soil test P exceeds the critical level, utilization or crop use of residual P reserves is recommended, with regular monitoring of soil test P levels to maintain critical soil test P. Using composite or field-average soil sampling protocols will not distinguish or separate P-responsive areas and excessive P areas within the field. Therefore, it is crucial to expand adoption of geospatial soil sampling technologies to quantify the spatial distribution of soil test P and spatially distribute fertilizer or manure P accordingly. Variable P application results in significant reductions in the spatial variation in soil test P, improving P efficiency.

Once the 'correct' P rates needed to meet crop P demand are quantified, other P management factors must be incorporated to further enhance P efficiency. For example, on P-responsive sites band-applied P generally increases P efficiency, especially when applied before or at planting. Although yield responses to inoculated seed or other additives to enhance P availability or solubilization of soil P residues have been tested, their effectiveness will be site-specific as with any P management decision (e.g. rate, source, timing, and placement).

Without question, conserving limited RP reserves for future generations will require enhanced P efficiency. The greatest challenge is associated with intensive animal production systems where manure P is annually applied in excess of crop P demand. Because of limited low soil test P land areas within an economically viable distance from the manure sources, it is imperative that alternative manure distribution technology be advanced to effectively utilize manure P to satisfy crop P demands, while minimizing off-site P transport to surface- and groundwaters.

7 Where to look for further information

Understanding nutrient use efficiency in agriculture requires a basic understanding of soil and crop management principles and practices related to plant nutrients. There are many textbooks available that provide a thorough foundation for the science behind efficient nutrient management. In the references cited we listed:

- Havlin, J. L., Tisdale, S. L., Nelson, W. L. and Beaton, J. D. 2014. *Soil Fertility and Fertilizers*.
- Syers, J. K., Johnston, A. and Curtin, D. 2008. Efficiency of soil and fertilizer phosphorus use: reconciling changing concepts of soil phosphorus behavior with agronomic information. FAO Fert. Plant Nutr. Bull. Vol. 18. FAO, Rome.

For more advanced summaries see the following book titles:

- Hawkesford, M. J., Kopriva, S. and de Kok, L. J. (Eds). 2014. *Nutrient Use Efficiency in Plants*. Springer International Publishing, Switzerland. ISBN: 978-3-319-10635-9.
- Rakshit, A., Singh, H. B. and Sen, A. (Eds). 2015. *Nutrient Use Efficiency: From Basics to Advances*. Springer-Verlag. ISBN 978-9-788-132-22168-5.
- Ramesh, K., Biswas, A., Lakaria, B. L., Srivastava, S. and Patra, A. K. (Eds). 2017. *Enhancing Nutrient Use Efficiency: Concepts, Methods And Management Interventions*. New India Publishing Agency, New Delhi, India. ISBN: 978-9-385-51673-3.

Current nutrient use efficiency research is commonly communicated by numerous international organizations and societies including:

- European Confederation of Soil Science Societies (ECSSS) (https://soilscience.eu/).
- Food and Agriculture Organization of the United Nations (FAO) (http://www.fao.org/).
- International Fertilizer Development Center (IFDC) (https://www.ifdc.org).
- Soil Science Society of America (SSSA) (https://www.soils.org/).

8 References

Abit, M. J. M., Arnall, D. B. and Phillips, S. B. 2018. Environmental implications of precision agriculture. In: Shannon, D. K., Clay, D. E. and Kitchen, N. R. (Eds), *Precision Agriculture Basics*. ASA, CSSA and SSSA, Madison, WI, pp. 209–20. doi:10.2134/precisionagbasics.2017.0035.

Alam, M., Bell, R., Salahin, N., Pathan, S., Mondol, A. T. M. A. I., Alam, M. J., Rashid, M. H., Paul, P. L. C., Hossain, M. I. and Shil, N. C. 2018. Banding of fertilizer improves phosphorus acquisition and yield of zero tillage maize by concentrating phosphorus in surface soil. *Sustainability* 10(9), 3234. doi:10.3390/su10093234.

An, G.-H., Kobayashi, S., Enoki, H., Sonobe, K., Muraki, M., Karasawa, T. and Ezawa, T. 2010. How does arbuscular mycorrhizal colonization vary with host plant genotype? An example based on maize (*Zea mays*) germplasms. *Plant Soil* 327(1–2), 441–53. doi:10.1007/s11104-009-0073-3.

Ao, J., Fu, J., Tian, J. and Liao, H. 2010. Genetic variability for root morph-architecture traits and root growth dynamics as related to phosphorus efficiency in soybean. *Funct. Plant Biol.* 37, 304–12.

Badri, D. V. and Vivanco, J. M. 2009. Regulation and function of root exudates. *Plant Cell Environ.* 32(6), 666–81. doi:10.1111/j.1365-3040.2008.01926.x.

Balaban, N., Suleimanova, A., Valeeva, L., Chastukhina, I., Rudakova, N., Sharipova, M. and Shakirov, E. 2017. Microbial phytases and phytate: exploring opportunities for sustainable phosphorus management in agriculture. *Amer. J. Mol. Biol.* 7, 11–29.

Barbieri, P. A., Rozas, H. R., Covacevich, F. and Echeverria, H. E. 2014. Phosphorus placement effects on phosphorus recovery efficiency and grain yield of wheat under no-tillage in the humid Pampas of Argentina. *Intl. J. Agron.* 2014, 1–12. doi:10.1155/2014/507105.

Batjes, N. H. 1997. A world dataset of derived soil properties by FAO-UNESCO soil unit for global modelling. *Soil Use Manag.* 13(1), 9–16. doi:10.1111/j.1475-2743.1997.tb00550.x.

Beauchemin, S., Hesterberg, D., Chou, J., Beauchemin, M., Simard, R. R. and Sayers, D. E. 2003. Speciation of phosphorus in phosphorus-enriched agricultural soils in X-ray absorption near-edge structure spectroscopy and chemical fractionation. *J. Environ. Qual.* 32(5), 1809–19.

Bly, A., Reicks, G. and Gelderman, R. 2015. Starter, banding, and broadcasting phosphorus fertilizer for profitable corn production. In Clay, D. E., Carlson, C. G., Clay, S. A. and Byamukama, E. (Eds), *IGrow Corn: Best Management Practices*. South Dakota State University, Brookings SD. Chapter 26.

Bovill, W. D., Huang, C. Y. and McDonald, G. K. 2013. Genetic approaches to enhancing phosphorus use efficiency (PUE) in crops: challenges and directions. *Crop Pasture Sci.* 64(3), 179–98. doi:10.1071/CP13135.

CAST. 2013. Food, fuel, and plant nutrient use in the future. Issue Paper No. 51. Council Agric. Sci. Tech., Ames, IA.

Chien, S. H., Edmeades, D., McBride, R. and Sahrawat, K. L. 2014. Review of maleic-itaconic acid copolymer purported as urease inhibitor and phosphorus enhancer in soils. *Agron. J.* 106(2), 423–30. doi:10.2134/agronj2013.0214.

Ciampitti, I. A. and Vyn, T. J. 2014. Understanding global and historical nutrient use efficiencies for closing maize yield gaps. *Agron. J.* 106(6), 2107–17. doi:10.2134/agronj14.0025.

Condron, L. M. and Newman, S. 2011. Revisiting the fundamentals of phosphorous fractionation of sediments and soil. *J. Soils Sediments* 11(5), 830–40. doi:10.1007/s11368-011-0363-2.

Cordell, D. and White, S. 2013. Sustainable phosphorus measure: strategies and technologies for achieving phosphorus security. *Agronomy* 3(1), 86–116. doi:10.3390/agronomy3010086.

Cordell, D., Drangert, J. and White, S. 2009. The story of phosphorus: global food security and food for thought. *Glob. Environ. Change* 19(2), 292–305. doi:10.1016/j. gloenvcha.2008.10.009.

Couto, R., Ferreira, P. A. A., Ceretta, C. A., Lourenzi, C. R., Facco, D. B., Tassinari, A., Piccin, R., De Conti, L. D., Gatiboni, L. C., Schapanski, D. and Brunetto, G. 2017. Phosphorus fractions in soil with a long history of organic waste and mineral fertilizer addition. *Bragantia* 76(1), 155–66. doi:10.1590/1678-4499.006.

Crozier, C. R., Havlin, J. L., Hoyt, G. D., Rideout, J. W. and McDaniel, R. 2009. Three experimental systems to evaluate phosphorus supply from enhanced granulated manure ash. *Agron. J.* 101(4), 880–8. doi:10.2134/agronj2008.0187x.

Delhaize, E. 2001. The role of root exudates in aluminum tolerance. In: Ae, N., Arihara, J., Okada, K. and Srinivasan, A. (Eds), *Plant Nutrient Acquisition*. Springer, Tokyo, Japan, pp. 140–55.

Dhillon, J., Torres, G., Driver, E., Figueiredo, B. and Raun, W. 2017. World phosphorus use efficiency in cereal crops. *Agron. J.* 109, 1–8.

Djinadou, K. A., Pierzynski, G. M. and Havlin, J. L. 1995. Phosphorus and micronutrient availability from dual application of nitrogen and phosphorus using liquid fertilizers. *Soil Sci.* 159(1), 49–58. doi:10.1097/00010694-199501000-00006.

Doydora, S., Hesterberg, D. and Klysubun, W. 2017. Phosphate solubilization from poorly crystalline iron and aluminum hydroxides by AVAIL copolymer. *Soil Sci. Soc. Am. J.* 81(1), 20–8. doi:10.2136/sssaj2016.08.0247.

Fixen, P. E. and Johnston, A. M. 2012. World fertilizer nutrient reserves: a view to the future. *J. Sci. Food Agric.* 92(5), 1001–5. doi:10.1002/jsfa.4532.

Fohse, D. and Jungk, A. 1983. Influence of phosphate and nitrate supply on root hair formation of rape, spinach and tomato plants. *Plant Soil* 74(3), 359–68. doi:10.1007/BF02181353.

Fuentes-Ramirez, L. E. 2005. Bacterial bio fertilizers. In: Siddiqui, Z. A. (Ed.), *PGPR: Bio-Control and Biofertilization*. Springer, the Netherlands, pp. 143–72.

Grant, C. A., Flaten, D. N., Tomasiewicz, D. J. and Sheppard, S. C. 2001. The importance of early season phosphorus nutrition. *Can. J. Plant Sci.* 81(2), 211–24. doi:10.4141/P00-093.

Greiner, R. 2007. Phytate-degrading enzymes: regulation of synthesis in microorganisms and plants. In: Turner, B. L. and Mullaney, E. J. (Eds), *Inositol Phosphate-Linking Agriculture and the Environment*. CABI, Wallingford, UK, pp. 78–96.

Havlin, J. L. and Heiniger, R. W. 2009. Variable fertilizer application decision support tool. *Precis. Agric.* 10(4), 356–69. doi:10.1007/s11119-009-9121-5.

Havlin, J. L., Tisdale, S. L., Nelson, W. L. and Beaton, J. D. 2014. *Soil Fertility and Nutrient Management: an Introduction to Nutrient Management* (8th edn.). Pearson, Upper Saddle River, NJ, 516pp.

Heathwaite, A., Sharpley, A., Beckmann, M. and Rekolainen, S. 2005. Assessing the risks and magnitude of agricultural nonpoint source phosphorus pollution. In: Sims, J.

T. and Sharpley, A. N. (Eds), *Phosphorus: Agriculture and the Environment*. Agron. Monogr. 46. ASA, CSSA and SSSA, Madison, WI.

Hedley, M. and McLaughlin, M. 2005. Reactions of phosphate fertilizers and by-products in soils. In: Sims, J. T. and Sharpley, A. N. (Eds), *Phosphorus: Agriculture and the Environment*. Agron. Monogr. 46. ASA, CSSA and SSSA, Madison, WI.

Hedley, M., Stewart, J. and Chauhan, B. 1982. Changes in the inorganic and organic soil phosphorus fractions induced by cultivation practices and by laboratory incubation. *Soil Sci. Soc. Amer. J.* 46, 970–6.

Heffer, P., Prud'homme, M., Muirheid, B. and Isherwood, K. 2006. Phosphorus fertilization: issues and outlook. Proceedings 586. International Fertilizer Society, London, UK, pp. 1–32.

Hendriks, L., Claassen, N. and Jungk, A. 1981. Phosphatverarmung des wurzelnahen bodens und phospataufnahme von mais und raps. *Z. Pflanzenernaehr. Bodenk.* 144(5), 486–99. doi:10.1002/jpln.19811440507.

Herring, J. R. and Fantel, R. J. 1993. Rock phosphate demand into the next century: impact on world food supply. *Nonrenew. Resour.* 2, 226–46.

Hesterberg, D. 2010. Macro-scale chemical properties and x-ray absorption spectroscopy of soil phosphorus. In: Singh, B. and Grafe, M. (Eds), *Applications of Synchrotron-Based Techniques in Soils and Sediments*. Elsevier, Burlington, MA, pp. 313–56.

Heuer, S., Gaxiola, R., Schilling, R., Herrera-Estrella, L., Lopez-Arrendondo, D., Wissuwa, M., Delhaize, E. and Rouached, H. 2017. Improving phosphorus use efficiency: a complex trait with emerging opportunities. *Plant J.* 90(5), 868–85. doi:10.1111/tpj.13423.

Hinsinger, P. 2001. Bioavailability of soil organic P in the rhizosphere as affected by root-induced chemical changes: a review. *Plant Soil* 237(2), 173–95. doi:10.1023/A:1013351617532.

Hopkins, B. G. 2013. Russet Burbank potato phosphorus fertilization with dicarboxylic acid copolymer additive (AVAIL®). *J. Plant Nutr.* 36(8), 1287–306. doi:10.1080/01904167.2013.785565.

Hossain, M. B. and Sattar, M. A. 2014. Effect of inorganic phosphorus fertilizer and inoculants on yield and phosphorus use efficiency of wheat. *J. Environ. Sci. Nat. Resour.* 7(1), 75–9. doi:10.3329/jesnr.v7i1.22148.

Idriss, E. E., Makarewicz, O., Farouk, A., Rosner, K., Greiner, R., Bochow, H., Richter, T. and Borriss, R. 2002. Extracellular phytase activity of *Bacillus amyloliquefaciens* FZB45 contributes to its plant-growth-promoting effect. *Microbiol.* 148(7), 2097–109. doi:10.1099/00221287-148-7-2097.

IFDC. 2010. *World Phosphate Rock Reserves and Resources*. International Fertilizer Development Center, Muscle Shoals, AL.

Isherwood, K. F. 2003. Fertilizer consumption and production: long term world prospects. Proceedings 507. International Fertilizer Society, York, UK, 23p.

Johnston, A. E., Poulton, P. R., Fixen, P. E. and Curtin, D. 2014. Phosphorus: its efficient use in agriculture. *Adv. Agron.* 123, 177–228.

Karpinets, T., Greenwood, D. and Ammons, J. 2004. Predictive mechanistic model of soil phosphorus dynamics with readily available inputs. *Soil Sci. Soc. Amer. J.* 86, 644–53.

Kellogg, R., Lander, C., Moffitt, D. and Gollehon, N. 2000. Manure nutrients relative to the capacity of cropland and pastureland to assimilate nutrients: spatial and temporal trends for the United States. U.S. Dept. of Agriculture, Natural Resources Conservation Service: Economic Research Service, Washington DC.

Kleinman, P. J. A., Sharpley, A. N., Buda, A. R., Easton, Z. M., Lory, J. A., Osmond, D. L., Radcliffe, D. E., Nelson, N. O., Veith, T. L. and Doody, D. G. 2017. The promise, practice, and state of planning tools to assess site vulnerability to runoff phosphorus loss. *J. Environ. Qual.* 46(6), 1243-9. doi:10.2134/jeq2017.10.0395.

Koide, R. T. and Kabir, Z. 2000. Extraradical hyphae of the mycorrhizal fungus *Glomus intraradices* can hydrolyze organic phosphate. *New Phytol.* 148(3), 511-7. doi:10.1046/j.1469-8137.2000.00776.x.

Korkmaz, K., Ibriki, H., Karnez, E., Buyuk, G., Ryan, J., Ulger, A. C. and Oguz, H. 2009. Phosphorus use efficiency of wheat genotypes grown in calcareous soils. *J. Plant Nutr.* 32(12), 2094-106. doi:10.1080/01904160903308176.

Kovar, J. L. 2001. The role of roots in maximum soil productivity. *Fluid J.* 9, 1-4.

Kovar, J. L. and Barber, S. A. 1987. Placing phosphorus and potassium for greatest recovery. *J. Fert. Issues* 4, 1-6.

Kovar, J. L. and Barber, S. A. 1988. Phosphorus supply characteristics of 33 soils as influenced by seven rates of phosphorus addition. *Soil Sci. Soc. Amer. J.* 52(1), 160-5. doi:10.2136/sssaj1988.03615995005200010028x.

Kovar, J. L. and Barber, S. A. 1989. Reasons for differences among soils in placement of phosphorus for maximum predicted uptake. *Soil Sci. Soc. Amer. J.* 53(6), 1733-6. doi:10.2136/sssaj1989.03615995005300060021x.

Kovar, J. L. and Claassen, N. 2005. Soil-root interactions and phosphorus nutrition of plants. In: Sims, J. T. and Sharpley, A. N. (Eds), *Phosphorus: Agriculture and the Environment*. Agron. Monogr. 46. ASA, CSSA and SSSA, Madison, WI.

Krishnapriya, V. and Pandey, R. 2016. Root exudation index: screening organic acid exudation and phosphorus acquisition efficiency in soybean genotypes. *Crop Pasture Sci.* 67(10), 1096-110. doi:10.1071/CP15329.

Leigh, R. A. and Johnston, A. E. 1986. An investigation of the usefulness of phosphorus concentrations in tissue water as indicators of the phosphorus status of field grown spring barley. *J. Agric. Sci.* 107(2), 329-33. doi:10.1017/S0021859600087128.

Li, Z. G. and Zhang, H. Y. 2001. Application of microbial fertilizers in sustainable agriculture. *J. Crop Prod.* 3(1), 337-47. doi:10.1300/J144v03n01_28.

Li, B., Boiarkina, I., Young, B., Yu, W. and Singhal, N. 2018. Prediction of future phosphate rock: a demand model. *J. Envir. Inform.* 31, 41-53.

Lindsay, W. L. 1979. *Chemical Equilibrium in Soils*. John Wiley & Sons, New York, NY.

Lombi, E., Scheckel, K. G., Armstrong, R. D., Forrester, S., Cutler, J. N. and Paterson, D. 2006. Speciation and distribution of phosphorus in a fertilized soil. *Soil Sci. Soc. Amer. J.* 70(6), 2038-48. doi:10.2136/sssaj2006.0051.

Lynch, J. P. 2007. Roots of the second green revolution. *Aust. J. Bot.* 55(5), 493-512. doi:10.1071/BT06118.

Malecki, M., Mouazen, A., De Ketelacre, B., Ramon, H. and De Baerdemacker, J. 2008. On-the-go variable-rate phosphorus fertilization based on a visible and near-infrared soil sensor. *Biosys. Eng.* 99, 35-46.

Mariano, E. D., Jorge, R. A., Keltjens, W. G. and Menossi, M. 2005. Metabolism and root exudation of organic acid anions under aluminum stress. *Braz. J. Plant Physiol.* 17(1), 157-72. doi:10.1590/S1677-04202005000100013.

Marschner, H. 1998. Role of root growth, arbuscular mycorrhiza, and root exudates for the efficiency in nutrient acquisition. *Field Crops Res.* 56(1-2), 203-7. doi:10.1016/S0378-4290(97)00131-7.

Mazid, M. and Khan, T. A. 2014. Future of bio-fertilizers in Indian agriculture: an overview. *Inter. J. Agric. Food Res.* 3(3), 10–23. doi:10.24102/ijafr.v3i3.132.

McBeath, T. M., McLaughlin, M. J., Armstrong, R. D., Bell, M., Bolland, M. D. A., Conyers, M. K., Holloway, R. E. and Mason, S. D. 2007. Predicting the response of wheat (*Triticum aestivum* L.) to liquid and granular phosphorus fertilizers in Australian soils. *Aust. J. Soil Res.* 45(6), 448–58. doi:10.1071/SR07044.

McKenzie, R. H. and Roberts, T. L. 1990. Soil and fertilizer phosphorus update. In: *Proceedings of the Alberta Soil Science Workshop*, pp. 84–104.

McLaughlin, M. J., McBeath, T. M., Smernik, R., Stacey, S. P., Ajiboye, B. and Guppy, C. 2011. The chemical nature of P accumulation in agricultural soils-implications for fertilizer management and design: an Australian perspective. *Plant Soil* 349(1–2), 69–87. doi:10.1007/s11104-011-0907-7.

Medinski, T., Freese, D. and Reitz, T. 2018. Changes in soil phosphorus balance and phosphorus-use-efficiency under long-term fertilization conducted on agriculturally used chernozem in Germany. *Can. J. Soil Sci.* 98(4), 650–62. doi:10.1139/cjss-2018-0061.

Mendes, F. F., Guimaraes, L. J. M., Souza, J. C., Guimaraes, P. E. O., Magalhaes, J. V., Garcia, A. A. F., Parentoni, S. N. and Guimaraes, C. T. 2014. Genetic architecture of phosphorus use efficiency in tropical maize cultivated in low-P soil. *Crop Sci.* 54(4), 1530–8. doi:10.2135/cropsci2013.11.0755.

Menezes-Blackburn, D., Zhang, H., Sutter, M., Giles, C. D., Darch, T., George, T. S., Shand, C., Lumsdon, D., Blackwell, M., Wearing, C., Cooper, P., Wendler, R., Brown, L. and Haygarth, P. M. 2016. A holistic approach to understanding the desorption of phosphorus in soils. *Environ. Sci. Technol.* 50(7), 3371–81. doi:10.1021/acs.est.5b05395.

Mengel, D. B. and Barber, S. A. 1974. Rate of nutrient uptake per unit of corn root under field conditions. *Agron. J.* 66(3), 399–402. doi:10.2134/agronj1974.00021962006600030019x.

Motavalli, P. P. and Miles, R. J. 2002. Inorganic and organic soil phosphorus fractions after long-term animal manure and fertilizer applications. *Better Crops* 86, 20–3.

Neto, A. P., Favarin, J. L., Hammond, J. P., Tezotto, T. and Couto, H. T. 2016. Analysis of phosphorus use efficiency traits in Coffee genotypes reveals *Coffea arabica* and *Coffea canephora* have contrasting phosphorus uptake and utilization efficiencies. *Front. Plant Sci.* 7, 408. doi:10.3389/fpls.2016.00408.

Niu, Y. F., Chai, R. S., Jin, G. L., Wang, H., Tang, C. X. and Zhang, Y. S. 2013. Responses of root architecture development to low phosphorus availability: a review. *Ann. Bot.* 112(2), 391–408. doi:10.1093/aob/mcs285.

Nkebiwe, P. M., Weinmann, M., Bar-Tal, A. and Muller, T. 2016. Fertilizer placement to improve crop nutrient acquisition and yield: a review and meta-analysis. *Field Crops Res.* 196, 389–401. doi:10.1016/j.fcr.2016.07.018.

Oyeyiola, Y. and Omueti, J. 2016. Phosphorus uptake and use efficiency by cowpea in phosphocompost and chemical fertilizer treated nutrient degraded acid soils. *Agric. Res. Tech.* 1, 1–8.

Parentoni, S. and Junior, C. 2008. Phosphorus acquisition and internal utilization efficiency in tropical maize genotypes. *Pesq. Agropec. Bras.* 43, 893–901.

Peterson, G. A., Sanders, D. H., Grabouski, P. H. and Hooker, M. L. 1981. A new look at row and broadcast phosphate recommendations for winter wheat. *Agron. J.* 73(1), 13–7. doi:10.2134/agronj1981.00021962007300010004x.

Rahman, M. M., Salleh, M. A. M., Rashid, U., Ahsan, A., Hossain, M. M. and Ra, C. S. 2014. Production of slow release crystal fertilizer from wastewaters through struvite crystallization-A review. *Arab. J. Chem.* 7(1), 139-55. doi:10.1016/j.arabjc.2013.10.007.

Rehim, A., Hussain, M., Abid, M., Zia-Ul-Haq, M. and Ahmad, S. 2012. Phosphorus use efficiency of *Triticum aestivum* L. as affected by band placement of phosphorus and farmland manure on calcareous soils. *Pak. J. Bot.* 44, 1391-8.

Richardson, A. E., Hocking, P. J., Simpson, R. J. and George, T. S. 2009. Plant mechanisms to optimise access to soil phosphorus. *Crop Pasture Sci.* 60(2), 124-43. doi:10.1071/CP07125.

Richardson, A. E., Lynch, J. P., Ryan, P. R., Delhaize, E., Smith, F. A., Smith, S. E., Harvey, P. R., Ryan, M. H., Veneklaas, E. J., Lambers, H., Oberson, A., Culvenor, R. A. and Simpson, R. J. 2011. Plant and microbial strategies to improve the phosphorus efficiency of agriculture. *Plant Soil* 349(1-2), 121-56. doi:10.1007/s11104-011-0950-4.

Roberts, T. L. and Johnston, A. E. 2015. Phosphorus use efficiency and management in agriculture. *Resour. Cons. Recycl.* 105, 275-81. doi:10.1016/j.resconrec.2015.09.013.

Roberts, T. and Stewart, W. 2002. Inorganic phosphorus and potassium production and reserves. *Better Crops* 86, 6-7.

Rose, T. J. and Wissuwa, M. 2012. Rethinking internal phosphorus utilization efficiency: a new approach is needed to improve PUE in grain crops. *Adv. Agron.* 116, 185-217.

Sander, D. H., Penas, E. J. and Eghball, B. 1990. Residual effects of various phosphorus application methods on winter wheat and grain sorghum. *Soil Sci. Soc. Am. J.* 54(5), 1473-8. doi:10.2136/sssaj1990.03615995005400050043x.

Sattari, S. Z., Bouwmann, A. F., Giller, K. E. and van Ittersum, M. K. 2012. Residual soil phosphorus as the missing piece in the global phosphorus crisis puzzle. *Proc. Natl. Acad. Sci. U. S. A.* 109(16), 6348-53. doi:10.1073/pnas.1113675109.

Schepers, J., Schlemmer, M. and Ferguson, R. 2000. Site-specific consideration for managing phosphorus. *J. Environ. Qual.* 29, 125-30.

Schlegel, A. J. and Havlin, J. L. 2017. Corn grain yield and grain uptake from 50 years of nitrogen and phosphorus fertilization. *Agron. J.* 109, 1-8.

Schlegel, A. J. and Havlin, J. L. 2019. *Personal communication.*

Scholz, R. W., Roy, A. H. and Hellums, D. T. 2014. Sustainable phosphorus management: a transdisciplinary challenge. In: Scholz, R., Roy, A., Brand, F., Hellums, D. and Ulrich, A. (Eds), *Sustainable Phosphorus Management*. Springer, Dordrecht.

Scholz, R. W., Hellums, D. T. and Roy, A. A. 2015. Global sustainable phosphorus management: a transdisciplinary venture. *Current. Sci.* 108(7), 3-12.

Schroder, J., Cordell, D., Smit, A. and Rosemarin, A. 2010. Sustainable use of phosphorus. Rep. 357. Plant Research International, Wageningen, the Netherlands, 124pp.

Schutz, L., Gattinger, A., Meier, M., Müller, A., Boller, T., Mäder, P. and Mathimaran, N. 2017. Improving crop yield and nutrient use efficiency via biofertilization - a global meta-analysis. *Front. Plant Sci.* 8, 2204. doi:10.3389/fpls.2017.02204.

Selles, F., Campbell, C. A., Zentner, R. P., Curtin, D., James, D. C. and Basnyat, P. 2011. Phosphorus use efficiency and long-term trends in soil available phosphorus in wheat production systems with and without nitrogen fertilizer. *Can. J. Soil Sci.* 91(1), 39-52. doi:10.4141/cjss10049.

Shabnam, R. and Iqbal, M. 2016. Phosphorus use efficiency by wheat plants grown in an acidic soil. *Braz. J. Sci. Tech.* 3, 18-33.

Sharpley, A., Beegle, D., Bolster, C., Good, L., Joern, B., Ketterings, Q., Lory, J., Mikkelsen, R., Osmond, D. and Vadas, P. 2012. Phosphorus indices: why we need to take stock of how we are doing. *J. Environ. Qual.* 41(6), 1711–9. doi:10.2134/jeq2012.0040.

Simeonova, D. D., Wilson, M. M., Metcalf, W. W. and Schink, B. 2010. Identification and heterologous expression of genes involved in anaerobic dissimilatory phosphite oxidation by *Desulfotignum phosphitoxidans*. *J. Bacteriol.* 192(19), 5237–44. doi:10.1128/JB.00541-10.

Sims, J. T. and Kleinman, P. J. A. 2005. Managing agricultural phosphorus for environmental protection. In: Sims, J. T. and Sharpley, A. N. (Eds), *Phosphorus: Agriculture and the Environment*. Agron. Monogr. 46. ASA, CSSA and SSSA, Madison, WI.

Singh, B. and Satyanarayana, T. 2011. Microbial phytases in phosphorus acquisition and plant growth promotion. *Phys. Mol. Bio. Plants* 17(2), 93–103. doi:10.1007/s12298-011-0062-x.

Smil, V. 2000. Phosphorus in the environment: natural flows and human interferences. *Ann. Rev. Energy Environ.* 25, 33–88.

Smit, A., Bindrabin, P., Schroder, J., Conijin, J. and Meer, H. 2009. Phosphorus in agriculture: global resources, trends and developments. Rep. 282. Plant Research International, Wageningen, the Netherlands, 36pp.

Su, S. and Miller, M. H. 1993. Determination of the most efficient phosphorus placement for field-grown maize (*Zea mays* L.) in early growth stages. *Can. J. Soil Sci.* 73(3), 349–58. doi:10.4141/cjss93-037.

Sun, M., Alikhani, J., Massoudieh, A., Greiner, R. and Jaisi, D. P. 2017. Phytate degradation by different phosphohydrolase enzymes: contrasting kinetics, decay rates, pathways, and isotope effects. *Soil Sci. Soc. Amer. J.* 81(1), 61–75. doi:10.2136/sssaj2016.07.0219.

Sweeney, D. W., Moyer, J. L. and Havlin, J. L. 1996. Multinutrient fertilization and placement to improve yield and nutrient concentration of tall fescue. *Agron. J.* 88(6), 982–6. doi:10.2134/agronj1996.00021962003600060023x.

Syers, J. K., Johnston, A. and Curtin, D. 2008. Efficiency of soil and fertilizer phosphorus use: reconciling changing concepts of soil phosphorus behavior with agronomic information. FAO fertilizer and plant nutrition bulletin 18. FAO, Rome.

Talboys, P. J., Heppell, J., Roose, T., Healey, J. R., Jones, D. L. and Withers, P. J. 2016. Struvite: a slow-release fertiliser for sustainable phosphorus management? *Plant Soil* 401, 109–23. doi:10.1007/s11104-015-2747-3.

USDA-NRCS. 2011. Conservation practices standard. Nutrient Management 590. USDA Natural Resources Conservation Service, Washington DC. Available at: https://www.nrcs.usda.gov/wps/portal/nrcs/main/national/landuse/crops/npm/ (accessed on 15 November 2018).

Vance, C. P., Uhde-Stone, C. and Allen, D. L. 2003. Phosphorus acquisition and use: critical adaptations by plants for securing a nonrenewable resource. *New Phytol.* 157(3), 423–47. doi:10.1046/j.1469-8137.2003.00695.x.

van der Wiel, C. C. M., van der Linden, C. G. and Scholten, O. E. 2016. Improving phosphorus use efficiency in agriculture: opportunities for breeding. *Euphytica* 207(1), 1–22. doi:10.1007/s10681-015-1572-3.

van Dijk, W. 2003. Adviesbasis voor de bemesting van akkerbouw- en vollegrondsgroentegewassen. PPO-Publicatie Praktijkonderzoek Plant & Omgeving nr. 307, Lelystad, 66pp.

Vaneeckhaute, C., Janda, J., Vanrolleghem, P. A., Tack, F. M. G. and Meers, E. 2016. Phosphorus use efficiency in bio-based fertilizers: a bio-availability and fractionation study. *Pedosphere* 26(3), 310–25. doi:10.1016/S1002-0160(15)60045-5.

van Kauwenbergh, S. 2010. World phosphate rock reserves and resources. IFDC Tech. Bull. 75. International Fertilizer Development Center, Muscle Shoals, AL.

Van Vuuren, D. P., Bouwman, A. F. and Beusen, A. H. W. 2010. Phosphorus demand for the 1970-2100 period: a scenario analysis of resource depletion. *Glob. Environ. Change* 20(3), 428–39. doi:10.1016/j.gloenvcha.2010.04.004.

Veneklass, E. J., Lambers, H., Bragg, J., Finnegan, P. M., Lovelock, C. E., Plaxton, W. C., Price, C. A., Scheible, W. R., Shane, M. W., White, P. J. and Raven, J. A. 2012. Opportunities for improving phosphorus use efficiency in crop plants. *New Phytol.* 195(2), 306–20. doi:10.1111/j.1469-8137.2012.04190.x.

Vidon, P. G., Welsh, M. K. and Hassanzadeh, Y. T. 2018. Twenty years of riparian zone research (1997–2017): where to next? *J. Environ. Qual.* 48(2), 248–60, doi:10.2134/jeq2018.01.0009.

Wang, L., Liao, H., Yan, X., Zhuang, B. and Dong, Y. 2004. Genetic variability for root hair traits as related to phosphorus status in soybean. *Plant Soil* 261(1/2), 77–84. doi:10.1023/B:PLSO.0000035552.94249.6a.

Wang, X., Shen, J. and Liao, H. 2010a. Acquisition or utilization, which is more critical for enhancing phosphorus efficiency in modern crops? *Plant Sci.* 179(4), 302–6. doi:10.1016/j.plantsci.2010.06.007.

Wang, X., Yan, X. and Liao, H. 2010b. Genetic improvement for phosphorus efficiency in soybean: a radical approach. *Ann. Bot.* 106(1), 215–22. doi:10.1093/aob/mcq029.

Webb, J. R., Mallarino, A. P. and Blackmer, A. M. 1992. Effects of residual and annually applied phosphorus on soil test values and yields of corn and soybean. *J. Prod. Agric.* 5(1), 148–52. doi:10.2134/jpa1992.0148.

Wissuwa, M., Katsuhiko, K., Fukuda, T., Mori, A., Rose, M. T., Pariasca-Tanaka, J., Kretzschmar, T., Haefele, S. M. and Rose, T. J. 2015. Unmasking novel loci for internal phosphorus utilization efficiency in rice germplasm through genome-wide association analysis. *PLoS ONE* 10(4), e0124215. doi:10.1371/journal.pone.0124215.

Wittry, D. J. and Mallarino, A. P. 2004. Comparison of uniform- and variable-rate phosphorus fertilization for corn-soybean rotations. *Agron. J.* 96(1), 26–33. doi:10.2134/agronj2004.0026.

Yan, X., Liao, H., Beebe, S. E., Blair, M. W. and Lynch, J. P. 2004. QTL mapping of root hair and acid exudation traits and their relationship to phosphorus uptake in common bean. *Plant Soil* 265(1-2), 17–29. doi:10.1007/s11104-005-0693-1.

Yetilmezsoy, K., Illhan, F., Kocak, E. and Akbin, H. M. 2017. Feasibility of struvite recovery process for fertilizer industry: a study of financial and economic analysis. *J. Cleaner Prod.* 152, 88–102. doi:10.1016/j.jclepro.2017.03.106.

Chapter 3

Delivering improved phosphorus acquisition by root systems in pasture and arable crops

Richard J. Simpson and Rebecca E. Haling, CSIRO Agriculture and Food, Australia; and Phillip Graham, Graham Advisory, Australia

1 Introduction

Intensification of food production since the 1940s has been underpinned by fertiliser use and is critical to our ability to feed the human population of the world (Stewart et al., 2005; Erisman et al., 2008). Phosphorus (P) is one of the handful of nutrients used in fertilisers that are now essential for global food security (Stewart et al., 2005; Sutton et al., 2013; Sattari et al., 2014). Most mineral P fertilisers are currently manufactured from phosphate rock, a non-renewable resource. Unfortunately, the global use of P is marked by inefficiencies in handling and use along the entire supply-demand chain (Weaver and Wong, 2011; Scholz and Wellmer, 2015; Scholz and Wellmer, 2018; Lun et al., 2018). In agriculture, only about half of the total P inputs to soil are recovered by crops. Roughly half of the unrecovered P is accumulated in fertilised soils, and

http://dx.doi.org/10.19103/AS.2020.0075.26

the other half becomes dispersed in the natural environment as a result of erosion, runoff or leaching (e.g. Vaccari et al., 2019; Lun et al., 2018; Bouwman et al., 2013; MacDonald et al., 2011; Cordell et al., 2009). Phosphorus lost from agriculture to the aquatic P cycle is contributing to eutrophication of waterways and ecosystem damage on an unprecedented scale throughout the world (Jarvie et al., 2013a,b; Campbell et al., 2017). Accelerated loss of P to the aquatic cycle also effectively ends our capacity to recycle this P for use in food production because the timeframe for P in the aquatic cycle is millennia (Filippelli, 2008; Föllmi, 1996).

Global estimates of P flows, such as these, clearly have their uses but they mask the diversity of farming systems across the world and, consequently, the opportunities for more sustainable use of the world's finite P reserves. There could not be a wider divergence in the P status of the soils that are used globally for agriculture. At one extreme are soils that are naturally P fertile or have been fertilised for so long (e.g. Vitousek et al., 2009; Sattari et al., 2012; Barrow and Debnath, 2014) that maximum yields can be achieved without continuous application of P fertiliser. Applications of P fertiliser to soils such as these, often exacerbate P losses to aquatic systems and contribute significantly to ecosystem damage (Jarvie et al., 2013b; Rowe et al., 2016). In contrast, vast areas of agricultural soil are P deficient for optimum crop growth, and production per hectare is either low, relative to the potential of the agricultural production system, or relies on the use of P fertiliser to achieve high yields (e.g. Stewart et al., 2005; Vitousek et al., 2009; Lynch, 2011).

In natural ecosystems, plants adapt to low-P soils using a variety of P acquisition mechanisms (e.g. acclimation of root architecture and morphology, mycorrhizal symbioses, specialised root structures, organic anion exudation, rhizosphere acidification), that either enhance the diffusion of P to the root or enable greater P interception and/or extraction (Lambers et al., 2008; Lambers et al., 2010; Richardson et al., 2011). Plants that evolve in these systems will often combine acclimation to low-P soil, with some form of P-use strategy that also confers a degree of low-P tolerance (e.g. slow growth, high internal P utilisation efficiency; Lambers et al., 2010; Waddell et al., 2015). However, sustainable, high crop growth rates are an imperative in an agricultural context because the social, environmental, and monetary costs of using land and water resources inefficiently for food and fibre production are too high (Tilman et al., 2011; Fischer and Connor, 2018).

It is in this context that agronomic practices can be used to enhance P acquisition by roots, and P-efficient plants can be used to produce more food and fibre from P-deficient soils, or to lower the fertiliser cost of correcting P deficiency for high production. Phosphorus-efficient plants can also help to mitigate the risk of P losses from agriculture to the wider environment.

2 Options for improving phosphorus acquisition by roots in pastures and arable crops

Phosphate in soil solution can be acquired by plants as a result of contact-exchange between the root and the soil (root interception), via mass flow in the movement of soil water towards the root, and by diffusion in the water-filled path between soil particles and the root. The relative contribution to P uptake by each mode of acquisition will vary depending on the concentration of available P in the soil. However, it is usual to expect that only 1% or less of available P will be supplied by root interception and 2–3% via mass flow due to the very low concentrations of phosphate in soil solution. The majority of P uptake occurs as a result of phosphate diffusion to the root (Barber, 1995; Tinker and Nye, 2000).

Diffusion of phosphate in soil may be approximated using Fick's first law of diffusion in which it is assumed that the rate of diffusion is stable over the time interval of measurement:

$$\mathbf{J} = -\mathbf{D} * \mathbf{A} * d\mathbf{C}/d\mathbf{x} \tag{1}$$

where: $\mathbf{J}$ is the diffusive flux (µmol/s), $\mathbf{A}$ is the area of the absorbing root-soil interface (cm^2), $\mathbf{D}$ is the isothermic diffusion coefficient (cm^2/s), $\mathbf{C}$ is the phosphate concentration (µmol/cm^3) and $d\mathbf{C}/d\mathbf{x}$ is the concentration gradient created between the root-soil interface and the bulk soil when the concentration of phosphate at the root-soil interface is depleted as P is taken up by the root.

The rate of phosphate diffusion in soil is very low compared to other nutrients (e.g. nitrate or potassium) and is rate limiting for P acquisition from soils up to the point where soil P fertility is sufficient for maximum growth rates (Barber, 1995; Tinker and Nye, 2000). Consequently, successful agronomic and plant-based strategies to increase P acquisition by plant roots have been guided predominantly by the key principles embodied in the diffusion equation, rather than the possibility of modifying other steps in P uptake and metabolism by plants (e.g. rates at which P is transported across cell membranes during P uptake from soil solution: Mitsukawa et al., 1997; Rae et al., 2004; Richardson, 2009; Heuer et al., 2017).

The key principles are: (1) the rate of P diffusion to the root is directly proportional to the P concentration gradient between the bulk soil and the absorptive surface of the root system, and is (2) inversely proportional to the distance over which P diffusion occurs; and (3) the rate of P uptake by the plant is directly determined by the size of the absorptive surface area of the root system or, in the case of plants colonised by arbuscular mycorrhizal fungi (AMF), the effective surface area of the root-hyphae system.

2.1 Managing phosphate diffusion

The phosphate concentration in soil solution (**C**, Eqn 1) is most effectively manipulated by the application of P fertilisers (particularly soluble P fertilisers) to directly increase the solution P concentration, but it may also be improved by modification of a plant's rhizosphere (see comments below). Fertiliser efficiency is often greatly improved when P is banded close to where roots are expected to grow. In cropping systems this is typically achieved by placing P fertiliser below, or to the side of seeds at planting. However, in a non-arable system such as a permanent pasture, a relatively concentrated band of P may also be created by broadcasting P fertiliser onto the surface of the soil. The phosphorus chemistry of banded P is complex (Kar et al., 2012) and the effectiveness of bands is influenced by the P buffering capacity of the soil to which they have been applied (Dibb et al., 1990). In some instances, the way the fertiliser band is created must also take account of soil chemistry to ensure that the potential benefits are realised (e.g. compare the effectiveness of granular-P and fluid-P fertiliser bands in calcareous soils, Lombi et al., 2004).

Nevertheless, the basic principles are that: (i) P banding allows **C** to be elevated in the vicinity of the root at a lower rate of P application than would otherwise be necessary; and (ii) the effectiveness of the P band is enhanced because plants usually increase **A** (the absorptive surface area of their root system, Eqn 1) by proliferating root length in the P band (Duncan and Ohlrogge, 1958; Yao and Barber, 1986). Savings in P fertiliser can be very large (e.g. 30–60% less P fertiliser, Jarvis and Bolland, 1991; up to 75% less P fertiliser, Sanchez et al., 1991) compared with broadcast applications of P fertiliser, but the relative benefit depends on the initial P fertility of the soil. The greatest benefit from banding P is achieved in P-deficient soil and the least benefit in soils that are already close to the soil P fertility level needed for maximum crop yield (Sanchez et al., 1991). The volume of a fertiliser band may also limit its effectiveness for P acquisition if it is too small, and therefore limits the root length proliferation that can be achieved within the band (i.e. increase in **A**). If it is necessary to apply higher rates of P, a larger band volume will be required to maximise P acquisition (e.g. Borkert and Barber, 1985; Duncan and Ohlrogge, 1958).

The diffusion coefficient and the distance over which P must diffuse (**D** and **dx**, Eqn 1) are primarily characteristics of the soil, but **dx** is strongly modified in drying soil because the tortuosity of the diffusion path is increased or the path is broken. Phosphorus concentrated near the soil surface is most vulnerable to soil drying and this accounts for numerous reports that deeper placement of P bands can substantially delay the impact of moisture stress on crop or pasture nutrition and growth (Scott, 1973; Cornish and Myers, 1977; Pinkerton and Simpson, 1986; Valizadeh et al., 2003)

2.2 Increasing the absorptive surface of a plant root system

Most plant-based examples of improved P-acquisition efficiency involve modification of **A** (the area of the P absorbing root-soil interface, Eqn 1). Increasing **A**, particularly in P-enriched zones of a P-deficient soil profile: (i) directly improves the acquisition of P (Lynch, 2019) and (ii) enables the plant to achieve its maximum yield at a lower value of **C** (Haling et al., 2016a; Haling et al., 2016b; Sandral et al., 2018). The former is important when the principal objective is to achieve higher yields in P-impoverished soils and low-input agriculture (e.g. Burridge et al., 2019); the latter is the key to reducing fertiliser input costs associated with P accumulation in fertilised soil or P loss by leaching and runoff (Simpson et al., 2014).

A variety of root system traits contribute to a large **A** for P acquisition (e.g. Table 1). Plants allocate mass to nutrient foraging roots, root mass is converted to length via the specific root length, and root length is translated into **A** via the root radius and root hair length. Phosphorus uptake by genotypes within a plant species is often correlated with total root length alone (i.e. root length is a reasonable proxy for **A**), but when comparing among species it becomes essential to also account for root radius and root hair length (e.g. McLachlan et al., 2019). It is, consequently, feasible to use the specific surface area of the root hair cylinder (specific SARHC = π * [root radius + root hair length]^2 * root length / root mass) as a first approximation for predicting which plant species are likely to have a lower 'critical'[1] P requirement and an improved capacity to acquire P from low-P soil (Yang et al., 2017; Sandral et al., 2018). However, specific SARHC alone may not predict P-efficient genotypes within a species because the acclimation of root traits to low-P soil (root plasticity) becomes increasingly important for improved P-acquisition efficiency within a species (e.g. Haling et al., 2016b; Zhu et al., 2010). The most successful strategy, to date, for identifying and developing crop genotypes and alternative plant species for improved P-acquisition efficiency in agriculture has been to utilise variation in nutrient foraging root traits (e.g. Table 1; Lynch, 2007; Lynch, 2011; Lynch, 2019; Burridge et al., 2019; Wissuwa et al., 2016; Gamuyao et al., 2012; Mori et al., 2016; Hufnagel et al., 2014; Manske et al., 2000; Zhao et al., 2004; Sandral et al., 2018; Sandral et al., 2019).

When a plant is colonised by AMF, **A** is also modified by the 'effective' surface for P acquisition of the mycorrhizal hyphae. However, the net effectiveness of the root-AMF association for P uptake may not be as large as anticipated from the potential reach of the hyphal network. In addition, selection for nutrient-foraging root traits appears to deliver improved

---

1 Here we use the term 'critical' P requirement to denote the P supply rate or soil test P concentration that supports near maximum crop yield in a P deficient soil. Typically, this is defined as the soil test P concentration corresponding with 90% or 95% of maximum yield (e.g. Gourley at al., 2019).

Table 1 Examples of roots traits that increase the root-soil interface and/or the efficiency of this interface for improved P acquisition from low-P soil.

Root trait	Impact/ benefits	Example(s)	Reference
Preferential allocation of mass to foraging roots (i.e. root proliferation)	Increased size of root system; topsoil foraging	Universal response of many species to nutrient deficiency including many temperate pasture legume species (e.g. *T. subterraneum*)	Brouwer (1962); Freschet et al. (2015); Haling et al. (2016a); Haling et al. (2018); Sandral et al. (2018)
Increased root number, branching, root length (e.g. basal, crown, lateral roots; high topsoil root length density)	Topsoil foraging	*Phaseolus vulgaris* L.; *Zea mays* L.; *Oryza sativa* L.; *Sorghum bicolor* (L.) Moench; *Triticum aestivum* L.	Miguel et al. (2013); Sun et al. (2018); Jia et al. (2018); Zhu and Lynch (2004); Gamuyao et al. (2012); Hufnagel et al. (2014); Manske et al. (2000)
Shallow root angles	Topsoil foraging	*Phaseolus vulgaris*; *Zea mays*; *Oryza sativa*; *Glycine max* (L.) Merr.	Miguel et al. (2015); Ho et al. (2005); Lynch (2011); Mori et al. (2016); Zhao et al. (2004)
High specific root length	Effective allocation of carbon to increased nutrient foraging;	*Ornithopus* spp.; *Trifolium subterraneum* L.; *Citrus sinensis* (L.) Osbeck	Haling et al. (2016a, 2018); Eissenstat (1991)
Long, dense root hairs	Effective allocation of carbon to increase root-soil interface	*Phaseolus vulgaris*; *Triticum aestivum* L.; *Hordeum vulgare* L.; *Oryza sativa*.; *Glycine max*; *Zea mays*	Miguel et al. (2015); Gahoonia et al. (1997); Nestler and Wissuwa (2016); Vandamme et al. (2013); Zhu et al. (2010)
Root cortical aerenchyma	Lower root metabolic costs	*Zea mays* (modelling)	Postma and Lynch (2011)
Root cortical senescence	Lower root metabolic costs	*Hordeum vulgare* (modelling); *Phaseolus vulgaris*	Schneider et al. (2017a); Strock et al. (2018)
Tolerance of biotic or abiotic constraints to root and/ or root hair growth	Increased size of root system/root-soil interface	*Hordeum vulgare* (Al-tolerance)	Delhaize et al. (2009)

P-acquisition efficiency irrespective of a plant's interactions with AMF (Strock et al., 2018; Lynch, 2019; McLachlan et al., 2020). Possible reasons for this are discussed below.

There is considerable diversity among plant species in the expression of P-efficient root traits, in root acclimation responses to localised P supply, and in the relative importance of the morphological and physiological responses of roots to P deficiency (e.g. Lyu et al., 2016). Consequently, a successful root trait strategy for improving P acquisition in one crop species may not always be ideal for improving P efficiency in another. For example, shallow axial root angles that increase the density of root length in topsoil layers have proven to be a useful way to improve P-acquisition efficiency in *Phaseolus vulgaris* L., *Glycine max* (L.) Merr. and *Zea mays* L. (Miguel et al., 2015; Zhao et al., 2004; Lynch, 2011) but were found to be unrelated to P-acquisition efficiency in *Sorghum bicolor* (L.) Moench (Marcus, 2013). Likewise, root length proliferation in a nutrient-rich patch of soil is a very common nutrient acquisition response (Robinson and Van Vuuren, 1998) yet some important species have relatively weak root proliferation responses to localised P supply (e.g. *Vicia faba* L., *Glycine max,* Li et al., 2014a; Lyu et al., 2016). Root hairs are often assumed to assist P acquisition by becoming longer in low-P soil (e.g. *Solanum lycopersicum* L., *Brassica napus* L., *Spinacia oleracea* L., *Zea mays* and *Arabidopsis thaliana* (L.) Heynh., Föhse and Jungk, 1983; Bates and Lynch, 1996; Zhu et al., 2010), but this is not the case in all species (e.g. *Trifolium* spp., Haling et al., 2016b; McLachlan et al., 2019; McLachlan et al., 2020). In contrast, flexible acclimation of root hairs to low-P soil exists among lines of *Zea mays*; root hairs may be relatively short with the ability to lengthen in low P conditions, short and unresponsive to P, or long and unresponsive to P (Zhu et al., 2010).

Combining beneficial P efficiency traits can have a synergistic impact on P acquisition. Miguel et al. (2015) demonstrated, using recombinant inbred lines of *P. vulgaris*, that varieties with long root hairs achieved 1.9-2.5-fold larger shoot yields than short root hair varieties and that shallow rooted varieties grew 1.6-2.1-fold larger than deep rooting varieties as a result of improved topsoil foraging and P acquisition in P deficient soil. However, varieties that combined long root hairs with shallow topsoil roots achieved the largest yields. The P acquisition benefit was larger than would be anticipated from the additive effect of these traits alone. However, it is unwise to assume that this will always be the case. Experiments that have compared the P acquisition of 'short' and 'long' root hair mutants (e.g. the *rht2* mutant of *Zea mays*, Wen and Schnable, 1994; or *RSL2* and *RSL3* mutants of *Brachypodium distachyon* (L.) P. Beauv., Zhang et al., 2018) have not always found longer, more dense root hairs to be beneficial (e.g. Zhang et al., 2018; Klamer et al., 2019). While adverse pleotropic effects of ectopic transgenes cannot be ruled out in these experiments, it is also observed that plants can be very plastic in their root system acclimation to P stress and may react to changes in one root trait by adjusting acclimation to P stress in another. This has been found to be the case in a short root hair variant of *Zea mays* which produces more fine roots than the long root hair variant and partially compensates for its shorter root hair length (Weber et al., 2018; Klamer et al., 2019). Consequently, traits that should confer improved P acquisition may not always appear to be associated with a net P-efficiency benefit (e.g. Nestler and Wissuwa, 2016).

2.3 Mycorrhizal fungi

With a few notable exceptions, AMF colonise the roots of a majority of agricultural plant species (Smith and Read, 2008) where they form potentially symbiotic associations that increase the surface area for P uptake (i.e. the mycorrhizosphere) through the development of an extensive root-hyphal network. In P-deficient soils, the potential increase in soil volume explored can be very large (e.g. ~15-fold, Mai et al., 2019) but the increase in yield due to improved P acquisition is not as large as might be anticipated (Richardson et al., 2011; McLachlan et al., 2020). Indeed, there is an on-going debate in agriculture concerning the benefit or otherwise of AMF in high production systems (Ryan and Graham, 2018). The discussion is multifacetted, but key issues that influence discussion of the degree to which AMF improve P-acquisition by plants include: (i) AMF can improve P acquisition by plants growing in P-deficient soil, but the P-acquisition benefit diminishes when P fertiliser is applied and crop yields may even be marginally depressed at optimum P supply (Ryan and Graham, 2018); this creates a mismatch in the optimum P supply for the AMF association versus that for high crop production (Mai et al., 2018), (ii) some crop species are clearly more AMF-dependent than others (Schweiger et al., 1995; Smith and Read, 2008; Ryan and Graham, 2018), (iii) yield and P uptake responses to AMF colonisation are often relatively large in sterilised, P-deficient soil, but the P acquisition benefits can range from negligible to significant when experiments are conducted using non-sterile field soils (Svenningsen et al., 2018; Cruz-Paredes et al., 2019). This is thought to be a consequence of competition with, or suppression of, AMF by other soil microbiota (Svenningsen et al., 2018; Cruz-Paredes et al., 2019). In addition to these issues, *P. vulgaris* variants that developed longer topsoil roots as a consequence of increased root cortical senescence have been shown to be more P-efficient despite significantly impaired AMF colonisation (Strock et al., 2018). The role of AMF in P acquisition is discussed further in the case study in Section 2.4.

2.4 Alleviation of constraints to root growth

Finally, it is important to remember that other agronomic strategies can also have a significant role in improving the absorptive capacity for P acquisition by crop roots. Alleviation of constraints to root growth such as hard, dry, or toxic (e.g. saline, acid) soil conditions will concurrently improve nutrient acquisition and, in some circumstances, may prove to be a more cost-effective and durable way to increase P-acquisition efficiency.

A clear example of this occurs in very acidic soils where aluminium (Al) concentrations are elevated in soil solution and become toxic for root growth. This in turn, restricts nutrient foraging and P acquisition (amongst other problems for a crop). Delhaize et al. (2009) demonstrated how the incorporation of a gene for tolerance of toxic aluminium restored root growth by *Hordeum vulgare* L. in a very acid soil and provided a partial restoration of P-acquisition efficiency. Only partial restoration of P acquisition was achieved because the gene protected the growth of axial and lateral root length, but did not protect the growth of root hairs and rhizosheath development which were also sensitive to Al-toxicity (Haling et al., 2010a,b). The application of lime to raise soil pH and eliminate Al-toxicity, consequently, provided a more complete solution for both root growth and P efficiency.

When applying these principles in the field, it is also essential to develop a wholistic understanding of the constraint to production. For example, the depth to which soil acidity extends in acid soil profiles varies substantially. It is often difficult to physically and economically address deep soil acidity by lime incorporation (Scott et al., 1997; Scott and Coombes, 2006). In these circumstances, the most effective and durable solution for improving nutrient acquisition is to use Al-tolerant varieties in combination with lime application (Scott et al., 1997; Scott et al., 2001).

3 Rhizosphere modifications that improve phosphorus acquisition

The phosphate concentration of soil solution (C, Eqn 1) may also be increased by plants that have a capacity to modify their rhizosphere in ways that promote desorption, solubilisation, or mineralisation of P from sparingly available soil P pools. The main ways this may be achieved include:

  i Acidification of the rhizosphere by exudation of protons.
     Acidification of rhizosphere soil dissolves acid-soluble forms of P that are, otherwise, not readily available for plant uptake (Hedley et al., 1982; Li et al., 2003).
  ii Exudation of organic acid anions from roots.

Organic acid anions increase the plant-availability of P by competing with inorganic P (Pi) and organic P (Po) for adsorption sites on soil particles, and mobilise sparingly soluble and sorbed P via ligand-promoted dissolution of P that is complexed with aluminium and iron oxides or calcium (Ryan et al., 2001; Wang and Lambers, 2020). When present in high enough concentrations, organic anions may modify the soil P chemistry significantly, reducing the binding strength of adsorbed phosphate, decreasing the chemical affinity of Al^{3+} for phosphate; (Wang and Lambers, 2020) and increasing the diffusion coefficient (i.e. **D**, Eqn 1) of phosphate in the soil solution by orders of magnitude (Gerke, 1994). These impacts are modified by soil/rhizosphere pH and soil type (Oburger et al., 2011; Wang and Lambers, 2020).

iii Exudation of phosphatases.

Soils have considerable phosphatase activity, but activity is enhanced substantially in the rhizosphere where it is associated with depletion of soil-organic P (Tarafdar and Jungk, 1987; Gahoonia and Nielsen, 1992; Chen et al., 2002; George et al., 2002). Although these assessments do not distinguish between plant- and microbially derived phosphatase activities, a wide range of plants are known to exude phosphatase from their roots. When grown in sterile culture, many plants can hydrolyse a variety of organic P sources (e.g. glucose-1-P) for subsequent uptake and use. However, some forms of organic P common in soils, such as inositol phosphates, are relatively poorly utilised by plant roots (e.g. Hayes et al., 2000; Richardson et al., 2001). It was shown that this particular deficiency in root capability could be remedied by introducing a culture of soil microorganisms to the plant growth medium (Richardson et al., 2001); this highlights the importance of co-operativity between plant roots and soil microorganisms.

iv Stimulation of microbial activity (other than that of the AMF) in the rhizosphere.

Microorganisms are crucial to the cycling of Pi and Po in the detritusphere and the rhizosphere with significant proportions of total culturable bacteria and fungi reported to have Pi-solubilising/mobilising activity in laboratory media (e.g. Kucey et al., 1989; Bowen and Rovira, 1999; Schneider et al., 2017b). Microbial biomass and activity in the rhizosphere is many-fold greater than in bulk soil as a consequence of the release of mucilages, sugars, organic acid anions, and more complex carbon derived from root turnover and sloughed cells (Jones et al., 2009), and the amount of P held in the microbial biomass is considerably larger than that in the soil solution P pool (Seeling and Zasoski, 1993). Phosphorus is cycled rapidly and continuously through the microbial biomass (Oehl et al., 2001) and although it is often assumed that rhizosphere microbial activity provides a net P mineralisation (or solubilisation) benefit to plants, this is not guaranteed given the low P environment of the rhizosphere and will depend on the C:P ratio of the organic residues supporting microbial activity, the ability of the microbiome to solubilise P from the mineral phase, and the dynamics of microbial turnover (Richardson and Simpson, 2011; Meyer et al., 2019). Indeed, the stable microbial turnover of P from the soil solution Pi pool observed by Oehl et al. (2001) under 'steady-state' microbial conditions suggests that a net microbial release of P to the rhizosphere may only occur when the microbial biomass is in decline. As with other rhizosphere modifications, the high abundance of microorganisms and elevated rhizosphere enzyme activities are typically localised to within only a few millimetres of the root surface (Tarafdar and Jungk, 1987; Kandeler et al., 2001; Chen et al., 2002; Marschner et al., 2012).

3.1 Increasing phosphorus acquisition by rhizosphere management

With the exception of enhanced rhizosphere microbial activity, the mechanisms used by plants to modify P acquisition in the rhizosphere are typically enhanced as part of a plant species' general response to growth in P-deficient soil (Vance et al., 2003; Richardson et al., 2011; Wang and Lambers, 2020). Rhizosphere modifications typically influence the soil within only a few millimetres of the surface of the root/root hair cylinder (Marschner et al., 2012). Consequently, for many plant species, rhizosphere modifications are complimentary to, and may not replace the importance of a large root system for effective soil exploration (e.g. Mat Hassan et al., 2013; Mori et al., 2016). The limitations of the localised impact of rhizosphere modification are also recognised in intercropping research, because 'facilitated' P acquisition (where an intercropped species increases the size of the available P pool and facilitates improved P access for its neighbour) requires 'intimate intermingling' of roots by the two intercropped species (e.g. *Vicia faba* - *Zea mays* intercropped on alkaline soils, Li et al., 2003; Hinsinger et al., 2011). However, this is sometimes forgotten in P-acquisition research where it is often assumed that beneficial changes in rhizosphere biochemistry will translate to significantly improved P acquisition in the field. Confidence in the value of organic acid anion secretion from roots is provided by species like *Lupinus albus* L. which develops clusters of specialised rootlets with long root hairs that exude very high concentrations of citrate (Dinkelaker et al., 1989), sufficient to influence soil ~6 mm or more from the cluster root (Dessureault-Rompre et al., 2007). The formation of each cluster of rootlets is transient and other exudates (phosphatases, protons, secondary metabolites and microbial inhibitors) are released along with citrate from cluster roots. The rootlets are concentrated in small volumes of soil; a strategy that is obviously not conducive to 'nutrient foraging' and has, instead, been described as a specialised 'P mining' strategy (Lambers et al., 2006; Lambers et al., 2010). Phosphate is very effectively extracted from sparingly available and residual soil P pools by *L. albus* (e.g. Mat Hassan et al., 2013). When transferring P-acquisition aspirations for organic anion exudation to other crop species that do not form cluster roots, it is important to question whether the magnitude of organic anion release will be sufficient to mobilise sparingly available P (e.g. Ryan et al., 2014).

Although mobilisation of sparingly available P in the rhizosphere, as a result of rhizosphere phosphatase or organic anion release, has been demonstrated experimentally (e.g. enhanced rhizosphere phosphatase activity (Ye et al., 2018), organic acid anion release (Mat Hassan et al., 2013)), these studies are in contrast with other glasshouse and field experiments where differences in rhizosphere phosphatase activity, or organic anion release have not underpinned improvement, or explained differences in P acquisition by key crop species (e.g. George et al., 2008; Ryan et al., 2014).

There is a need for more wholistic studies that attempt to partition P acquisition quantitatively among all components of a plant's P acquisition strategy (e.g. root architecture, root morphology, rhizosphere modification) to guide plant improvement efforts along the most rewarding path as this is likely to be species-specific. Something akin to this has been attempted in experiments examining the root morphological and physiological responses of several crop species to low soil P (Mori et al., 2016; Lyu et al., 2016; Wang et al., 2016; Wen et al., 2019). Lyu et al. (2016) differentiated crop species on the basis of their primary reliance on root morphological plasticity (e.g. *Zea mays, Triticum aestivum* L.), as opposed to physiological plasticity (rhizosphere modification) (e.g. *Lupinus albus, Cicer arietinum* L.). Species differences were further defined by their reliance on rhizosphere phosphatase activity, organic anion release, and mycorrhizal colonisation (Wang et al., 2016; Wen et al., 2019). These approaches are a welcome improvement to assumptions that rhizosphere modifications are equally beneficial strategies for P acquisition, but they still do not quantify the proportions of P uptake that are due to the contrasting P acquisition strategies represented within a species (e.g. something similar to that attempted by Mori et al., 2016); this will be challenging to achieve.

4 Understanding the agronomic context in which improved phosphorus acquisition by roots can deliver benefits

The agricultural and environmental benefits of improving P acquisition by plant roots depend on the context in which improved P acquisition is achieved. It is important to evaluate the key elements of the P balance of a farming system (Fig. 1a) in relation to the P retention characteristics of the soil to understand how benefits in P efficiency may be achieved (Simpson et al., 2014). Perfect efficiency in the use of P fertiliser is achieved when P inputs to a farming system as fertiliser equal the removal of P in farm products. The metric we use to define this is referred to as the Phosphorus Balance Efficiency (PBE) where:

$$PBE\,(\%)=100*\left(\frac{P\ output\ in\ agricultural\ products}{P\ inputs\ in\ fertiliser\ and\ imported\ feed}\right) \qquad (2)$$

In a fertilised farm system, when P inputs are at, or approach equality with P outputs, this equation returns a value that approaches, or equals 100%. However, an anomaly of the calculations is that it is also possible to achieve, or exceed 100% in very low production systems that are not adequately fertilised because the output of P in products is depleting soil P reserves. This is clearly not a sustainable option. Here we confine discussion of the PBE metric to when it is applied to fertilised agriculture.

Johnston et al. (2014) describe P fertiliser application to crops grown in the UK and the USA, on soils with optimum P fertility, which achieve PBEs of 94–100%. By definition, these farming systems must have no P loss or P accumulation because P inputs equal P output in products (Fig. 1a). The 'P-sorption'[2] capacity of these soils must be low, with sufficient P retention within the soil profile to prevent leaching of P from the root zone. These systems would also have low soil loss and minimal P in runoff. In a system operating with near-perfect PBE, using plants with an improved capacity to acquire P from soil can have no beneficial impact on P use. Theoretically, P use in these farming systems can only be reduced by improving the internal P-utilisation efficiency of the crop in ways that reduce the amount of P exported from fields (e.g. Veneklaas et al., 2012).

4.1 Agricultural fields with low to very low phosphorus-sorbing soils

Very often low P-sorption capacity coincides with low P-buffering (e.g. sandy soils which have low phosphate-adsorption capacity and a poor ability to buffer changes in the P concentration of soil solution; Ozanne, 1980). This can result in large losses of P applied as fertiliser due to leaching and/ or runoff. Accumulation of P in these soils after fertiliser addition can be a relatively small component of the farming system's P budget (e.g. Russell, 1960; Ozanne et al.,

---

2 Here we use the term 'P sorption' to imply the net process of phosphate movement from soil solution to the solid phase of the soil and ultimately into forms of phosphate that are sparingly available for plant uptake as defined by Barrow (1999).

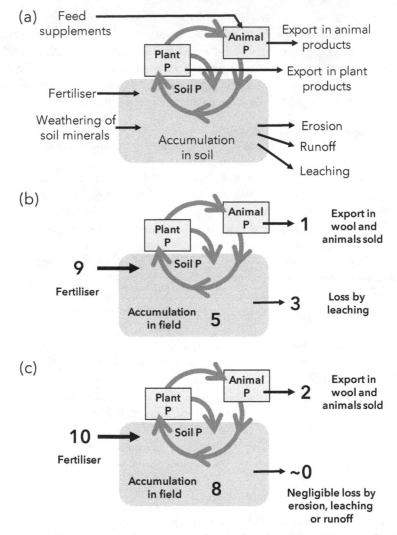

Figure 1 (a) The main inputs, outputs, and sinks for phosphorus (P) in a P-fertilised farming system. (b) Major net flows and accumulations of P in a permanent annual grass-annual legume pasture system growing on a low P-buffering, sandy soil near Willalooka, South Australia, that was grazed by sheep for wool and mutton production (data derived from Lewis et al., 1987). (c) Major net flows and accumulations of P in a permanent perennial grass-annual legume pasture growing on a soil with moderate P-buffering capacity (Phosphorus Buffering Index = 50; Burkitt et al., 2002) near Canberra, Australia. The pasture was grazed continuously by sheep for wool and sheep meat production and was being maintained at a soil test P concentration (0–10 cm depth) that was near the optimum for pasture growth (P1SR18 treatment from Simpson et al., 2015).

1961). For example, Fig. 1b shows the P budget of a fertilised, sheep grazing system in the Willalooka region of South Australia and demonstrates how P loss can contribute significantly to P inefficiency in a farm system based on sandy soils. The PBE of this sheep grazing system was only 13%; a third of the P applied as fertiliser was leached below the root zone each year and, of the P retained in the soil, about 55% was associated with soil organic matter (Lewis et al., 1987).

P losses of this nature can be substantially reduced by using P sources that are less soluble and release P slowly. However, there is often a trade-off between production and P loss because the ability of plants to access P from a less soluble P fertiliser is compromised (Ozanne et al., 1961). Using plant genotypes with lower external P requirements for maximum yield can also help mitigate the risk of P loss because the magnitude of loss via runoff and leaching is correlated with the soil test P (STP) concentrations of topsoil (Sharpley, 1995; McDowell et al., 2001; Melland et al., 2008; Bai et al., 2013; Withers et al., 2019). However, a benefit is only achieved when the use of these plants is combined with management of P inputs to reduce the soil's available-P concentration to the lower 'critical' STP requirement of the P-efficient plant. A strong nutrient foraging response, deeper rooted plants, and perennial species with deep, established root systems will also assist in capturing P that has been leached to depth (i.e. similar features to those of plants used to capture water and N that is leached to depth in a soil profile, Ridley et al., 1997; Dunbabin et al., 2003).

4.2 Agricultural fields with moderate to high phosphorus-sorbing soils

In contrast to soils with a low sorption capacity, soils with a moderate to high P-sorption capacity continue to slowly react with phosphate in ways that render it sparingly available for plant uptake (Barrow, 1974; Sample et al., 1980; Bolland, 1986a; Barrow, 1999; McLaughlin et al., 2011). In these soils, the rate of P diffusion to roots (and hence P acquisition) is improved by applying fertiliser to build the plant-available P concentration of the soil to the critical requirement of the crop or pasture (i.e. the STP concentration at which near-maximum yield is achieved). However, a component of the applied P inevitably accumulates in the soil as sparingly-available Pi. P application also results in the accumulation of Po (McLaughlin et al., 1990; Simpson et al., 2014; McLaren et al., 2020), only a proportion of which is immediately bio-available (Jarosch et al., 2015; Jarosch et al., 2019). Maintenance of stable soil-P fertility in these soils requires P inputs that cover P removal in agricultural products plus enough P to counteract the accumulations of P in the soil. Inevitably, budgets of P inputs and P outputs to fertilised agriculture on P-sorbing soils indicate that there is a P surplus. Indeed, management for optimum crop production (i.e. when soil fertility is maintained at the critical STP requirement of the crop) requires that there be a P surplus.

Figure 1c illustrates the key P flows in a southern Australian grassland system experiment managed with near-optimum soil P fertility, under continuous grazing. P accumulation in the field (mostly in sparingly-available pools of soil Pi and Po; McLaren et al., 2015) accounted for ~80% of the P fertiliser that it was necessary to apply annually to maintain near-maximum pasture yield (Simpson et al., 2015). Consequently, the PBE of the system was very low (20%).

Fertilising to cover P accumulations in high P-sorbing soils should not be confused with the problem of 'over-fertilising', when land is fertilised in excess of the P requirement for maximum crop yield (i.e. above the critical P requirement). Over-fertilising also results in a P budget surplus and low P-use efficiency (e.g. Simpson et al., 2015), but excessive accumulation of P in the soil is unwarranted and incurs an unnecessarily high risk of P loss to the environment (Gourley and Weaver, 2012; Jarvie et al., 2013a). Unfortunately, global budgets of P use in agriculture do not distinguish between P surpluses caused by 'best-practice' fertiliser use on P-sorbing soils, and excessive P-fertiliser use (e.g. MacDonald et al., 2011; Lun et al., 2018). Indeed, the tendency is to always assume that a P surplus represents poor fertiliser practice.

Excessive P application to soil must be recognised for what it is and should be addressed by reducing P inputs until critical soil P fertility levels for production and/or the environment are re-established (Gourley and Weaver, 2012).

Plants with root traits that enhance nutrient 'foraging' or 'mining' are associated with lower critical P requirements. In P-deficient soils they can deliver improved crop yields (Lynch, 2007; Lynch, 2019; Richardson et al., 2011) and in fertilised soils that accumulate P, the P inputs required for high production can be reduced (Simpson et al., 2011a, 2014).

5 Critical phosphorus benchmarks for soil phosphorus management

The most productive use of P fertiliser occurs when a P-deficient soil is fertilised to achieve the critical P requirement of the soil-plant system. Critical P specifications in 'amounts of P applied' are not suitable benchmarks for crop management because they are not transferable between soils that differ in their P fertility status or P buffering capacities. However, critical STP benchmarks are potentially transferable and are widely promoted (e.g. Gourley et al., 2019; Bell et al., 2013; Bai et al., 2013; Johnston et al., 2013; Poulton et al., 2013; Dodd and Mallarino, 2005). They are essential for effective management of crop nutrition, soil fertility, and the environmental risks associated with fertiliser use

(Rowe et al., 2016) provided they have been determined reliably and are used correctly.

Conceptually, a critical STP benchmark represents the available-P concentration of the soil that can support sufficient diffusion of P to the crop's root system for maximum growth. It is, by necessity, a rudimentary approximation of the complex interplay between a growing root system and diffusion of P to it. In practice the critical STP requirement of a crop is, therefore, influenced by many variables including plant species, stage of crop growth, soil type (unless P buffering capacity of the soil is accounted for e.g. Ozanne and Shaw, 1967; Gourley et al., 2019), soil sampling depth and depth distribution of P (Helyar and Price, 1999; Probert and Jones, 1977). Difficulty accounting for these factors during the empirical definition of a critical STP benchmark, or when making fertiliser decisions, can impact adversely on the perceived reliability of STP benchmarks. This may lead to doubt about their usefulness (e.g. Rowe et al., 2016), use that is restricted to the region in which a benchmark was determined, or fertiliser practice that reverts to unguided rules-of-thumb (e.g. as described in Reuter et al., 1995) with haphazard soil P fertility outcomes (e.g. see Fig. 2). In some instances, reliable critical nutrient benchmarks have been determined but they are ignored (e.g. Gourley et al., 2015). These scenarios can result in over-fertilisation (Sattari et al., 2012; Rowe

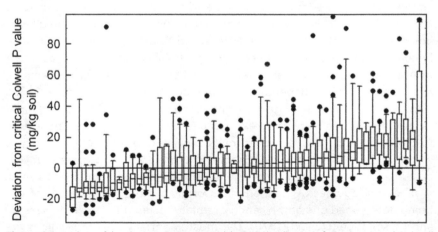

Figure 2 Box plots of the deviation in extractable P (Colwell, 1963) from the critical Colwell P concentration of soil in fields from 53 farms on the southern or central tablelands of New South Wales, Australia that were sampled (0–10 cm depth) in Spring 2016. Positive values indicate soil P fertility in the supra-optimal range; negative values indicate fields that were P-deficient. The graph shows the ranges in soil P fertility that existed on many farms prior to commencement of a regular soil fertility monitoring program. The box contains results from the middle 50% of fields, 80% of values occur inside the 'whiskers', and closed circles indicate fields with very high or very low STP concentrations. Bars show the median STP concentration. Redrawn from Simpson et al. (2017).

et al., 2016; Withers et al., 2019) which cannot be mitigated by improvements in plant P-acquisition efficiency.

6 Case study: improving the phosphorus efficiency of sheep and beef grassland farming in southern Australia

In this section we present a case study to demonstrate how diffusion of phosphate to roots in a grassland system can be managed to deliver effective use of P fertiliser in moderate to high P-sorbing soils. This was achieved by application of P banding technology, combined with annual monitoring of STP concentrations in the P 'band' to guide the pragmatic achievement of critical STP benchmarks for high grassland production. Improving knowledge of the P cycle in fertilised pastures and the factors that influence P sorption and accumulation in fertilised soil have shown how the introduction of novel plants with improved P-acquisition efficiency may enable a further substantive improvement in P-use efficiency in this grassland system.

6.1 Farming system context

Grassland farming is conducted on vast areas of P-deficient soil in the temperate zone of southern Australia (approximately delineated as south of latitude 32 S on the east coast to latitude 29 S on the west coast of Australia; Moore, 1970a). The climate is Mediterranean (winter-dominant rainfall with hot dry summers) in Western Australia and South Australia, with rainfall becoming more evenly distributed through the year towards the east coast of Australia. The soils are also deficient in N, and often other nutrients (S, K, and micronutrients depending on soil type; Williams and Andrew, 1970). Nitrogen (N) fertilisers are generally not used except in irrigated and high-rainfall, high-return dairy pastures, or occasionally as a strategic application to boost dryland pasture growth rates in winter. Instead, N deficiency is corrected by growing pasture legumes that are sown with grass in mixed pastures, or are oversown or encouraged to invade in natural grasslands. A variety of legumes of Mediterranean origin are used; their areas of use being determined primarily by annual rainfall (i.e. length of growing season) and by soil type (Donald, 1970). The most important pasture legumes are annual *Medicago* spp. (annual medics; low rainfall, neutral to alkaline soils), *Trifolium subterraneum* L. (subterranean clover; 300 mm – >900 mm annual rainfall, acid to neutral soils), *T. repens* L. (white clover; >900 mm rainfall, acid to neutral soils). *Medicago sativa* L. (lucerne or alfalfa) is used to a lesser extent as both a pasture and forage species across a wide climatic area provided soils are not subject to waterlogging or extreme acidity (Nichols et al., 2012).

Of these key legumes, *T. subterraneum* is the most widely used, with cultivars selected to fit differing growing-season lengths across the steep rainfall gradient of the temperate region in which it is grown (Donald, 1970; Nichols et al., 2013). This clover alone underpins the productivity of about 29.3 million hectares of dryland pasture (Hill and Donald, 1998), an area the size of Italy. The area of pasture production sown to *T. subterraneum* is the primary focus of this case study. Water (rainfall) is the key limiting resource in the grassland system. It determines total potential production and financial return per hectare; indeed, farms are often compared on the basis of production per 100 mm rainfall received (e.g. McEachern and Brown, 2010). High variability in rainfall contributes significantly to business risk and limits management options. In these environments, pastures are re-sown to improved varieties relatively infrequently. The high risk of a re-sown pasture failing to establish due to drought or because of low pasture renovation skill, combined with long periods (5-9 years) to repay the capital cost of pasture improvement are major disincentives (Lewis et al., 2012; Jackson and Malcolm, 2018). As a result, pastures are essentially permanent and are anticipated to remain in production for up to 20 years (Malcolm et al., 2014). New pasture varieties are selected for production combined with persistence reflecting this reality (e.g. Culvenor and Simpson, 2016).

Low soil fertility limits the efficiency of rainfall use (Mills et al., 2006) and application of P fertiliser is the easiest and most economical means to achieve high land-use efficiency and realise the productive potential of most farms (Carter and Day, 1970; Curll, 1977; Cayley et al., 1999; Osman et al., 1991; Smith et al., 2012; Lean et al., 1997). However, the PBE of sheep and beef production systems is very low and farmers need to apply 5- to 9-fold more P as fertiliser than they remove in animal products (median PBE = 9-19%; Weaver and Wong, 2011). This is due mainly to the moderate to high P-sorbing soils of the region which accumulate large proportions of the P that is applied as fertiliser (e.g. McLaughlin et al., 2011; Weaver and Wong, 2011; Simpson et al., 2015). Consequently, P fertiliser is a significant production cost for dryland grazing farms. It typically accounts for 20-25% of annual variable costs and is often the largest cost after labour and debt-servicing (McEachern and Brown, 2010).

6.2 Fertiliser practices to improve phosphorus acquisition on grassland farms

The most effective practices that can be employed to improve P uptake from these intrinsically P-deficient soils relate to the management of P diffusion to roots and include: use of soluble P fertilisers; localised banding of P at high concentrations within the root zone; and development and maintenance of an optimum soil available-P concentration. Here we outline these practices and explain the practical issues that define their use on farms.

6.2.1 Soluble phosphorus fertilisers

To achieve a rapid plant growth response when fertilising P-deficient soil, it is necessary to increase the phosphate concentration of the soil solution. This is most effectively done by applying soluble forms of P fertiliser to the soil (most often as single superphosphate in these systems). Experiments demonstrate repeatably that soluble P fertilisers deliver the most rapid increases in pasture production, and the largest yield gains per unit P applied (e.g. Sale et al., 1997; Lewis et al., 1997; Leech et al., 2019). Even when comparing among different P fertiliser products (including manures), their relative performance can usually be predicted by knowing their soluble P content (Leech et al., 2019). However, in the right circumstances there can be a role for less soluble P fertilisers (e.g. 'highly reactive' rock phosphate on very acid soils (Sale et al., 1997) and, particularly, on soils prone to P leaching (Lewis et al., 1997)).

6.2.2 Phosphorus banding

Subsurface banding of P that increases the concentration of phosphate close to a developing root system is a proven way of improving P fertiliser efficiency in crop systems (Jarvis and Bolland, 1991; Sanchez et al., 1991; see Section 2.1 'Managing phosphate diffusion'). It has also been shown, experimentally, to offer some advantages in pasture systems when surface soils become dry (Scott, 1973; Cornish and Myers, 1977; Pinkerton and Simpson, 1986), but it is neither practical nor economic to create subsurface P bands in permanent pastures. Nevertheless, P is usually relatively concentrated in the topsoil layer under pasture because the soil is subject to minimal disturbance and P in excreta and leaf-litter is deposited on the soil surface. This 'natural' surface P band is further enhanced by broadcasting soluble P fertiliser onto the soil surface in granular form.

When applied in a granule, phosphate moves rapidly from the granule into the soil, laterally and vertically, forming a P-enriched hemisphere below the granule (Fig. 3, Benbi and Gilkes, 1987; Hedley and McLaughlin, 2005). The concentration of phosphate in soil solution is highest near the granule and can exceed the P adsorption capacity of the soil (precipitates may form). P buffering is reduced in the high P zone and the high solution phosphate concentrations encourage diffusion of P into the P banded zone. In contrast, when a low solubility P source is applied to soil, the P concentration of the soil solution remains relatively low; this does not assist P diffusion to roots to the same extent (Lewis et al., 1997; Hedley and McLaughlin, 2005). The extent and speed of P movement into the soil is influenced by the P buffering

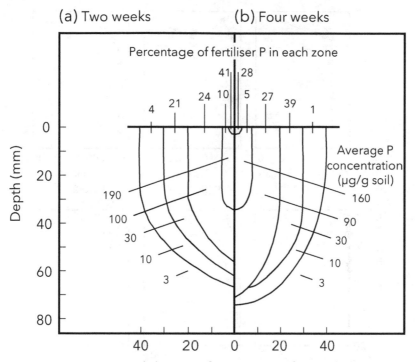

(a) Two weeks   **(b) Four weeks**

Percentage of fertiliser P in each zone

Figure 3 Distribution of phosphorus (P) in the soil (a podzolised sand) adjacent to a 135 mg granule of triple superphosphate (19.9% total P; 19.1% citrate-soluble; 16.0% water-soluble) at two and four weeks after its application to the soil surface. Adapted by permission from Springer Nature Customer Service Centre GmbH: Springer, Fertilizer Research, The movement into soil of P from superphosphate grains and its availability to plants, Benbi and Gilkes (1987).

capacity of the soil (Benbi and Gilkes, 1987). The dimensions of the band are initially shaped by the diffusion of P from P-fertiliser granules placed on the soil surface. However, empirical evidence indicates that ~70–80% of the increase in plant-available P after fertiliser application occurs in the topmost 10 cm of most soil profiles; the remaining impact is usually measured in the 10–20 cm layer of the topsoil even after a century of fertiliser use (Fig. 4; McCaskill and Cayley, 2000; Simpson et al., 2015; Schefe et al., 2015). Pasture species preferentially proliferate root length in the P-enriched surface soil band (Denton et al., 2006; Haling et al., 2016a,b). Heavy grazing, which can reduce total root production and root length at depth, does not diminish nutrient foraging in the surface soil layer (e.g. Greenwood and Hutchinson, 1998).

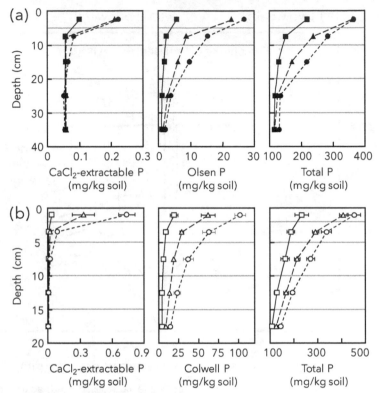

Figure 4 (a) Extractable P (0.01 M CaCl$_2$), Olsen extractable P (Olsen et al., 1954), and total P (nitric-perchloric acid digest) profiles of topsoil in the Permanent Top-Dressed pasture experiment at the Rutherglen Research Station, Rutherglen, Australia established as an unreplicated demonstration of the value of superphosphate applications in 1914 (Schefe et al., 2015). The data show soil P profiles in the unfertilised field (closed squares), a field to which 125 kg single superphosphate (~8.8% P) was applied per hectare every second year (closed triangles) and a field to which 250 kg single superphosphate was applied per hectare every second year (closed circles). The soil was sampled at 0-5, 5-10, 10-20, 20-30, and 30-40 cm depth intervals in November 2013 after 100 years of fertiliser use. The Phosphorus Buffering Index (PBI; Burkitt et al., 2008) of the top 10 cm of the P-fertilised soil profiles was ~115 (redrawn from Schefe et al., 2015). (b) Extractable P (0.005 M CaCl$_2$), Colwell extractable P (Colwell, 1963), and total P (sulfuric acid digest) profiles of topsoil from three soil P management treatments in a long-term grazed pasture experiment commenced near Canberra, Australia, in 1994 (Simpson et al., 2015). The soil P treatments were: unfertilised except for P inputs in supplementary feed for sheep (total P input over 20 year experiment = 5 kg P/ha; open squares), near-optimal soil P management (276 kg P/ha/20 years; open triangles) and supra-optimal P management for pasture growth (390 kg P/ha/20 years; open circles). The soil was sampled at 0-2, 2-5, 5-10, 10-15, and 15-20 cm depths in October 2014. Bars represent ± standard error; n= 3 replicate fields. The PBI of the top 10 cm of the soil profile was 50 (R. J. Simpson et al., unpublished data).

6.2.3 Development and maintenance of optimum phosphorus fertility

The plant-available P concentrations of surface P bands are managed by annual STP monitoring that is guided by empirically-derived STP benchmarks for high pasture production. Many of the practical problems that can prevent widespread application of soil test benchmarks to guide fertiliser management are overcome or avoided in the following ways:

i The most widely used pasture legumes in southern Australia have similar, relatively high, critical STP requirements compared to the grasses with which they are grown (Ozanne et al., 1969; Helyar and Anderson, 1970; Hill et al., 2005; Sandral et al., 2019). Atmospheric N fixed by the legume is the dominant source of N for growth by the pasture system. The critical P requirement of a mixed pasture is determined, consequently, by its legume component and this determines the benchmark STP concentration for optimum soil P fertility and high pasture production (Gourley et al., 2019).

ii Shallow testing of topsoils (0–10 cm) directly assesses ~70–80% of the plant available P in soil profiles (i.e. the surface P band; Fig. 4)[3]. Long-term experiments have demonstrated that in soils that are moderately P buffered, there is little leaching of P down the soil profile even after a century of regular P fertiliser applications, or after high rates of P applied over shorter timeframes (McCaskill and Cayley, 2000; Schefe et al., 2015; Simpson et al., 2015).

iii The STP benchmarks for pasture production are defined as the topsoil (0–10 cm depth) concentration of extractable P that corresponds with 95% of maximum growth *in spring* (Gourley et al., 2019). At this time of the year, temperature and soil moisture conditions are conducive to rapid growth rates (e.g. Cullen et al., 2008) and the demand for P by the crop is at its greatest. Farmers are encouraged to test pasture soils in spring and respond to the tests with a P fertiliser application in autumn after the summer drought and close to the opening of the next season's rainfall. Testing in spring removes the need to predict how P will be released during the growing season and assesses soil P fertility at the time of maximum crop demand. For a pasture soil that is being maintained close to its critical STP concentration, the fertiliser application creates a transient spike in P availability (see Fig. 5a; Simpson

3 The assumption that available P is found mainly in the uppermost layer of the soil breaks down in light-textured soils with very low P-buffering capacity (e.g. PBI <35) because soil P leaches from the uppermost soil layer (e.g. Lewis et al., 1981; Ritchie and Weaver 1993). Under these circumstances critical topsoil nutrient benchmarks may not be reliable indicators of pasture yield potential; testing soil samples taken to an appropriate depth must be considered (e.g. Probert and Jones, 1977).

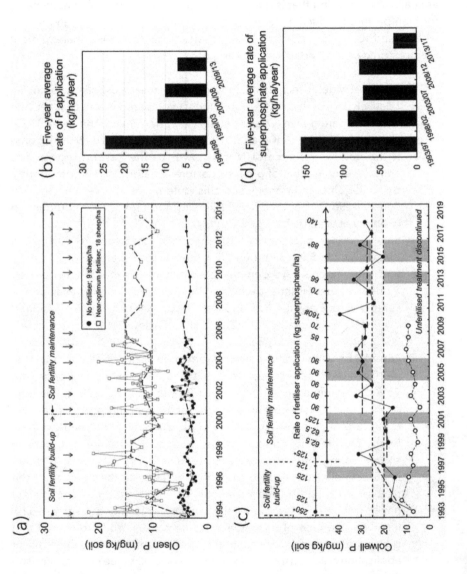

Figure 5 (a) The extractable P (Olsen et al., 1954) concentration of topsoil (0–10 cm) in selected grazing system treatments at Hall, ACT, Australia (adapted from Simpson et al., 2015). Three phases of the experiment are shown: 1994–2000, a soil fertility building phase, and 2001–2006, a soil fertility maintenance phase during which pasture was grazed continuously and STP was monitored at approximately 6-week intervals; and 2007–2014, a phase in which soil P fertility continued to be maintained, but with changed grazing management and only annual monitoring of the STP concentration. The fields that received no P fertiliser (closed circles) were stocked with 9 sheep/ha. Phosphorus was applied to the fertilised fields (open squares) with the intention of entering the spring period of pasture growth with a soil test P (STP) level within an Olsen P target band of 10–15 mg P/kg soil. These fields were stocked with 18 sheep/ha. Dashed horizontal lines delineate the target range for STP management. Arrows indicate when P fertiliser was applied. Each symbol represents the mean STP concentration of three replicate fields. Soil test P monitoring points are joined by a solid grey line to illustrate seasonal variability in soil test results. The dashed black line joins soil tests that were taken in Jan/Feb. These STP values were used to estimate the amount of P to apply each year. The dashed black line illustrates the variation in STP results that would be typical of annual monitoring at the same time each year. Phosphorus fertiliser was typically applied in autumn close to the break of each season. Error bars are not shown to improve the clarity of the figure. However, average coefficients of variation for the extractable P concentrations were 9.2%. (b) Five-year average rates of P application for the duration of the experiment. (c) Fertiliser application history and results of annual STP monitoring (Colwell, 1963) in a Grazing Systems Demonstration at, Bookham, NSW, Australia. Shaded vertical panels indicate years in which spring droughts occurred. The fertilised field carried 12–15 sheep/ha once the target STP concentration was achieved (zone delineated by dashed horizontal lines). Soil test results from an adjacent unfertilised paddock grazed continuously by 6 sheep/ha are also shown. The dashed lines show 'trends' in the data used at intervals to gauge progress in STP management. Sources: Graham (2006); Graham (2017); Simpson et al. (2009); R. P. Graham, unpublished data. (d) Five-year average rates of superphosphate (9% P) application for the duration of the Grazing Systems Demonstration.

et al., 2015) that exceeds the P requirement for early season growth and, provided the application rate for STP maintenance has been calculated correctly, the critical P level will be achieved at the peak of spring growth when demand for P is also peaking.

iv Because it is common for soils to have more than one potentially limiting nutrient (e.g. Trotter et al., 2014), successful application of the critical nutrient benchmarks also requires that there are no other limiting factors, or that they will be corrected concurrently. Critical soil test benchmarks for the most common secondary macronutrients limitations (e.g. S and K; Gourley et al., 2019) and soil toxicities (e.g. low pH/high Al; Slattery et al., 1999) have been determined and can be checked routinely.

6.3 Some pitfalls that have impeded progress

6.3.1 Confidence in the critical soil test phosphorus benchmarks for pasture production

It has long been recognised that optimum soil fertility management requires soil P fertility to be built by fertiliser application until the critical P benchmark for the soil-plant system is achieved, and for soil P management to transition to a maintenance phase during which the critical level of soil fertility is maintained (e.g. Fig. 5; McLachlan, 1965; Rudd, 1972; Holford and Crocker, 1988; Reuter et al., 1995). However, STP targets for fertiliser use were not generally promoted despite soil testing becoming generally accessible during the 1970s. This was partly because critical Colwell P concentrations (the most popular extractable P test; Colwell, 1963) were known to differ with soil type (e.g. Rudd, 1972). These issues were not resolved until 2007 (Gourley et al., 2007; Moody, 2007), when a major effort was made to reconcile historic data from ~650 fertiliser experiments conducted between 1955 and 2006 (Gourley et al., 2019). Earlier research had established the P requirement for maximum yield was related to the P buffering capacity of the soil (Ozanne and Shaw, 1967; Helyar and Spencer, 1977). It was demonstrated, using the historic data, that the critical Colwell P concentrations of pasture grown in different soil types could be predicted from the P buffering capacity of the soil. Concurrent testing for extractable P and Phosphorus Buffering Index (PBI) (a one-step test; Burkitt et al., 2002; Burkitt et al., 2008) was adopted almost immediately by the industry (Gourley et al., 2019).

6.3.2 Understanding and managing variability in soil test phosphorus concentrations

Targeted application of fertiliser P using a critical STP benchmark is illustrated in Fig. 5. There is always considerable seasonal variation in STP concentrations associated with time of fertiliser application (Fig. 5a), soil moisture/temperature conditions which stimulate bursts of P release from the microbial biomass, and P mineralisation/immobilisation as soils wet and dry (Perrott et al., 1990; Perrott et al., 1992; Blackwell et al., 2009). In addition, it is inevitable that sampling error, lack of replication in farm testing regimes, and error associated with the test itself will contribute to variability in soil test results. Before it was recognised that high variability in STP concentrations was eroding confidence in the use of soil tests, it was not uncommon for soil tests to be conducted at any time of the year, on an irregular basis and by irregularly using different P soil test methods. Farmers are now encouraged to test soils at the same time each year to minimise seasonal variability in test results, to set a pragmatic target range for soil P management that is guided by the critical P requirement of their soil-pasture system but is also cognisant of the intrinsic seasonal variability of soil test results, and to commence an annual soil-fertility monitoring program that will reveal temporal trends in soil P fertility (Simpson et al., 2009). As much as possible, fertiliser decisions are based on trends in the data (Fig. 5c), rather than one-off soil test results which may provide an unrealistic fertility assessment due to sampling and/or seasonal errors.

This approach is proven to maintain P supply to pasture roots for sustained high pasture and animal production and has been delivering a slow, but steady reduction in the fertiliser cost of production over time (Fig. 5b and c; Simpson et al., 2015). The reduction in costs can be attributed to at least two factors: (i) once the critical P fertility level is achieved, a lower fertiliser rate is required to maintain the target soil fertility than that required to build soil P fertility and this will deliver an obvious cost saving, and (ii) application of P to soil with moderate to high P buffering and sorption capacities has a 'P-sparing effect' (Barrow, 2015; Barrow et al., 2018). It is argued that the P-sparing arises because the negative charge on reactive surfaces in the soil is increased when sorbed P diffusively penetrates the soil particle; this reduces the P buffering capacity of the soil and the rate of further sorption reactions. Put simply, each application of P is expected to slowly improve the effectiveness of subsequent P applications (Bolland and Baker, 1998; Barrow et al., 1998, 2018).

In the examples provided (Fig. 5b and d), the initial step-change in P application rates was associated with the shift from building, to maintenance of soil P fertility. Thereafter, it is anticipated that P-sparing has contributed to the declining rate of maintenance P. However, it is difficult to determine the extent to which this was so, because Australian farmers also deal with a highly variable and drought-prone farming environment. Dry seasons are common and result in less P utilisation (less pasture growth), lower microbial activity and less P sorption; this means it is not necessary to apply as much fertiliser to maintain optimum fertility immediately after dry seasons. This can be seen in Fig. 5c where STP monitoring improved confidence in P fertiliser decisions and, increasingly, the decision was taken to skip fertiliser applications after dry seasons based on STP monitoring that indicated soil P-fertility had been conserved.

6.4 Progress towards the next step-change in the efficiency of phosphorus use for grassland production

Even with the use of soluble fertilisers, effective surface banding of fertiliser and good management of optimal P fertility, the PBE of grassland farms in southern Australia is low. Cropping systems in which soluble P fertiliser is banded with the seed can achieve median P balance efficiencies of ~48%. However, median P output in animal products from sheep and beef farms on similar soils is only 11–19% of P inputs (Weaver and Wong, 2011). Radiotracer experiments have demonstrated that the proportion of P captured by *T. subterraneum* pasture from a current superphosphate application is high (~42-50%; McLaren et al., 2017a) and is equivalent to, or better than that of crops (e.g. McLaughlin et al., 1988; McBeath et al., 2012). The P balance efficiency of grazed pasture is low because the pasture is not harvested like a crop. P removal in animal products is relatively small and most of the P acquired by the pasture is recycled (via the deposition of excreta, and death and decay of pasture) to the soil where it is re-exposed to P sorption/accumulation reactions (Donald and Williams, 1954; McLaughlin et al., 1990; Simpson et al., 2015; McLaren et al., 2017b). Direct phosphorus loss from these fields by leaching or runoff is relatively low (<1 kg/ha/year; Ridley et al., 2003) and does not account for the low PBE of the grassland system.

6.4.1 Factors that influence phosphorus accumulation in grazed fields

The accumulation of phosphate in acid soil can be mimicked in isothermic laboratory incubations and is described by the following empirical relationship:

$$Ps = a * C^{b1} * t^{b2} \tag{3}$$

where: **Ps** is the rate of phosphate sorption, **a** approximates the amount of sorbing material in a soil, **C** is the concentration of phosphate in soil solution, **t** is the time over which the reaction continues, and **b1** and **b2** are coefficients that describe the shape of the sorption relationship.

The **b1** and **b2** coefficients of Equation 3 vary widely between soils. However, they are reasonably well correlated when compared across a wide range of soils (Barrow, 1980a,b). Consequently, P sorption in the field is

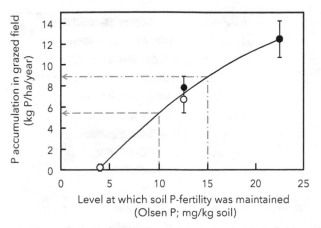

Figure 6 Average annual rates of phosphorus (P) accumulation in fields with a pasture based on *Trifolium subterraneum* (subterranean clover), *Phalaris aquatica* (phalaris) and annual grasses (e.g. *Bromus* spp., *Vulpia* spp.). The soil was acid (pH$_{Ca}$ 4.6) and had a moderate P-buffering capacity (P buffering index = 50, Burkitt et al. 2002). The pasture was grazed continuously by 9 sheep/ha (O) or 18 sheep/ha (●) and maintained at three levels of extractable P in the topsoil (0–10 cm depth) over 6 years (data derived from Simpson et al., 2015). Soil fertility levels are the midpoint soil test value (Olsen et al., 1954) of the target range for soil fertility management. The expected critical Olsen P concentration for near maximum pasture growth in this system was 15 mg P/kg. Bars represent ± standard error. The dash-dot line indicates the expected annual accumulation of P if the soil is maintained at the soil test P (STP) concentration recommended for *T. subterraneum*-based pasture; the dashed line indicates the expected accumulation of P if soil P fertility is maintained near the critical STP concentration of *Ornithopus* spp. (see Fig. 7 and Sandral et al., 2019). Redrawn from Simpson et al. (2014).

expected to be proportional to the plant-available phosphate concentration at which a soil is maintained.

However, P accumulation in a fertilised grazing system soil is more complex than this because it also involves accumulation of P in soil organic matter and to a lesser extent in very small areas of fields where excreta accumulates due to animal congregation behaviours. The proportions of inorganic P (Pi) and organic P (Po) accumulating in the soil vary depending on the P status of the soil. Typically, more Po is accumulated than Pi when a P-deficient pasture is fertilised and pasture growth is being increased; similar proportions of Po and Pi are accumulated when P levels are near optimal for pasture production, and Pi accumulation exceeds Po accumulation under conditions of supra-optimal P fertility (Simpson et al., 1974; McLaren et al., 2015; McLaren et al., 2020).

Results from a long-term, P fertility management experiment grazed by sheep have been used to assess whether management of plant-available

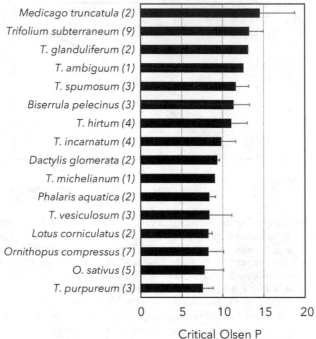

Figure 7 Average 'critical' extractable phosphorus (P) values (Olsen et al., 1954) for 14 pasture legumes and 2 grasses from a range of independent 'cultivar-site-year' assessments over three years and four field sites (Phosphorus Buffering Index range: 40-80; Burkitt et al., 2002). Critical soil test P values support 95% of maximum yield in spring. *Medicago sativa* was also included in these experiments but its critical P requirement could not be determined reliably because it often exceeded soil test P levels in the highest P treatments. Numbers in parentheses indicate the number of 'site-year' assessments made for each species. Bars representing 1x standard deviation are shown as a measure of the repeatability of the critical P determinations. Data were derived from Sandral et al. (2019).

phosphate was a dominant factor in the net accumulation of P in fertilised fields under grazing (Fig. 6; Simpson et al., 2014; Simpson et al., 2015). A positive (curvilinear) relationship between the STP target for soil P management and the annual rates of P accumulation in grazed fields was observed. Evidence from other long-term grazing experiments, while limited, supports this observation (Simpson et al., 2014). As described above, the experiment confirmed that maintaining soil P fertility close to the critical P requirement of the *T. subterraneum*-based pasture delivered the most effective use of P fertiliser for livestock production. Importantly, it also demonstrated that significantly higher rates of P fertiliser were needed to maintain supra-optimal STP concentrations because the rate of P accumulation in the soil was increased (Simpson et al.,

2015; McLaren et al., 2015, 2017b). This further underlined the folly of fertilising soils to levels well in excess of the critical P requirement of a farming system.

6.4.2 A clue to the development of a more phosphorus-efficient grazing system

The counterpoint to the previous observation was that using a productive, alternative pasture legume with a lower critical P requirement should enable the grazing system to be maintained at a lower STP concentration without impacting production adversely. This, in turn, would reduce P accumulation in grazed fields and less P fertiliser would be needed to maintain productivity (Fig. 6; Simpson et al., 2014).

The projected savings in the amount of P fertiliser necessary to maintain a lower critical STP concentration appeared to be substantial. For example, if it were possible to shift the target for soil P management from an Olsen P concentration (Olsen et al., 1954) of 15 mg P/kg soil (the optimum for *T. subterraneum* pastures: Gourley et al., 2019; Sandral et al., 2019) to 10 mg P/kg soil (i.e. close to the optimum subsequently found for serradellas (*Ornithopus* spp. – a P-efficient, alternative legume; Fig. 7, Sandral et al., 2019), the maintenance P cost of optimum pasture production would be reduced by about 30% (Simpson et al., 2014). To put this in context, the projected proportional reduction in the use of fertiliser P by this strategy is as large as the projected proportional benefit of recovering P from waste streams in Europe using novel and emerging technologies (17–31%; Tonini et al., 2019). Obviously, the quantum of the benefits in global terms and the degree-of-difficulty in implementing either strategy are probably not comparable.

6.4.3 The search for phosphorus-efficient pasture legumes

The task of finding alternative, P-efficient legume(s) to underpin improved P use efficiency was made potentially feasible by the fact that the P requirement of a single species of pasture legume (*T. subterraneum*) effectively determined the P requirement of this grassland farming system. However, the unparalleled success and wide adaptation of *T. subterraneum* also underlined the difficulty of the task. *Trifolium subterraneum* is grown in mixtures with various annual and perennial grasses and is well adapted to Mediterranean and temperate climates with highly variable rainfall and regular droughts. It is valued for its high productivity and persistence (Morley 1961; Rossiter, 1966; Nichols et al., 2012), feeding value for livestock (Doyle et al., 1993), N-fixation (Peoples et al., 2012), grazing tolerance (Nichols et al., 2012), and has now undergone about a century of selection and cultivar improvement (Nichols et al., 2013).

Much of the Australian flora have evolved in P-impoverished soils (Handreck, 1997) and this has led to interest in the suitability of native perennial legumes as an alternative to *T. subterraneum*. However, while there is a wide range in the critical P requirements among both native legumes and grasses, with some capable of reasonable growth in very P deficient media (Pang et al., 2010; Waddell et al., 2015), P-efficiency among the legumes was not accompanied by high potential yields in moderately fertile soils (e.g. Pang et al., 2010; Robinson et al., 2007). It is surmised that there is a potential role for these legumes, especially as ephemeral or opportunistic crops and pastures in arid and semi-arid environments (e.g. Nicol et al., 2013; Bell et al., 2012), but it is unlikely that they provide a viable P-efficient alternative to *T. subterraneum*.

Significant effort has also been expended to identify alternative pasture legumes from the Mediterranean basin as a way of diversifying the pasture base for animal production systems in southern Australia (Nichols et al., 2007; Nichols et al., 2012). The emphasis on legumes naturalised from the Mediterranean region reflects their successful acclimation to Australian soils and climate, their responsiveness to fertiliser applications, high herbage yields and seed production, and their persistence in managed grasslands (Donald, 1970). Nichols et al. (2013) detail 33 annual legume species registered for use in southern Australia in addition to *T. subterraneum*. Although 23 of these species were suited to acid soils and are, therefore, potential alternatives to *T. subterraneum*, 15 of the 23 species were also regarded as only suitable for localised use, rarely used, or of limited commercial value.

With this background in mind, Sandral et al. (2018) commenced a glasshouse experiment to determine whether there was any evidence for differing critical P requirements among 10 alternative annual legumes and *T. subterraneum*. Precautions were taken to mimic field conditions (e.g. the legumes were grown as micro-swards with P stratified in the topsoil to reflect surface P banding) to obtain realistic assessments of critical P requirements relative to *T. subterraneum*. However, final assessment of legumes truly capable of delivering lower critical P requirements was based on their subsequent growth responses to P fertiliser at four field sites under contrasting seasonal conditions (Fig. 7; Sandral et al., 2019). Nine of the ten species tested in the glasshouse were found to have a significantly lower critical P requirement than *T. subterraneum* (Sandral et al., 2018), but this was reduced to five legume species (out of 14 alternative legumes tested) that had consistently lower critical P requirements than *T. subterraneum* in field environments (Sandral et al., 2019).

Three of the legumes with lower critical P requirements were forage legumes species (crimson clover, *T. incarnatum* L.; purple clover, *T. purpureum* Loisel; arrowleaf clover, *T. vesiculosum* Savi). These species are suited to short-term pastures but have a limited role in the target farming zone where the high

Table 2 Critical P requirements (concentration of P applied to a topsoil layer (mg/kg soil) that corresponded with 90% of maximum yield), and comparative root morphology of nutrient foraging (topsoil) roots (mean ± s.e.) among three alternative pasture legumes measured at the soil P level (30 mg P applied/kg topsoil layer) where root proliferation by *T. subterraneum* was at its peak in a controlled-environment study of P acquisition (Haling et al., 2016a,b).

Parameter	subterranean clover cv. Leura (*Trifolium subterraneum*)	French serradella cv. Margurita (*Ornithopus sativus*)	yellow serradella cv. Santorini (*O. compressus*)
Critical P requirement (mg P/kg topsoil)	89 ± 1	38 ± 2	25 ± 2
Root mass (g/microsward)	0.61 ± 0.01	0.33 ± 0.01	0.27 ± 0.01
Root mass fraction (g/g)	0.24 ± 0.01	0.11 ± 0.004	0.08 ± 0.003
Specific root length (m/g root)	338 ± 23	459 ± 37	542 ± 26
Root length density (cm/cm³)	53 ± 4	35 ± 3	35 ± 1
Root hair length (mm)	0.23 ± 0.004	0.63 ± 0.03	0.58 ± 0.02
Surface area of root hair cylinder (cm²/microsward)	3359 ± 156	4686 ± 509	4245 ± 188

cost of re-sowing pastures carries a high financial risk relative to permanent pastures (Malcolm et al., 2014). However, soil P management recommendations for these P-efficient forage legumes can now be modified to guide fertiliser use in forage legume crops, during a pasture renovation sequence, or as a legume rotation phase in a cropping sequence. Only two pasture legume species with lower critical P requirements (*O. compressus* L., yellow serradella; *O. sativus* Brot., French (or pink) serradella) were considered potentially suitable for use in permanent pastures.

The field experiments demonstrated that all of the P-efficient alternatives could be grown with an Olsen STP concentration (0–10 cm) of 10 mg P/kg to achieve near-maximum spring yields in soils where it is recommended that *T. subterraneum* be grown with an Olsen STP concentration of 15 mg/kg (Sandral et al., 2019; Gourley et al., 2019). To date, our search within the *T. subterraneum* genome for P-efficient clover lines has not identified genotypes that can achieve the step-change in P-acquisition efficiency that is possible with the *Ornithopus* spp. (McLachlan et al., 2019).

6.4.4 How *Ornithopus* species achieve a lower critical phosphorus requirement

The critical P requirement among a large number of pasture legumes is determined by their ability to develop a large root-soil interface for P acquisition (Haling et al., 2016b; Sandral et al., 2018). In practice, a low critical P requirement has been associated with: (i) high capacity for proliferation of root mass in soil patches where P is relatively concentrated (e.g. in the topsoil P band created by broadcasting fertiliser), (ii) long specific root length, which enables root mass to be converted effectively into root length, and (iii) long root hairs, which convert root length into a large root cylinder volume. Comparisons of the root morphology and low-soil-P acclimation responses of *Ornithopus* spp. and *T. subterraneum* (Table 2; Haling et al., 2016b; Yang et al., 2017; Sandral et al., 2018) indicate that although the clover is very effective at proliferating root length in low-P soil, its relatively short specific root length means that this is only achieved by allocating a large amount of root dry mass to nutrient foraging. The proportion of plant mass allocated to nutrient foraging by *T. subterraneum* can be twice that allocated by an *Ornithopus* spp. when growing with a similar low P supply (e.g. Table 2). In addition, the clover has relatively short root hairs (0.23 mm) which prevent it from developing a comparable large root cylinder volume to that of the serradella with long root hairs (~0.6 mm), even when the clover has proliferated longer total length of nutrient foraging roots (e.g. Table 2; Haling et al., 2016a,b). Although other factors, such as exudation of organic anions, may play some role in P acquisition, comparative assessment of anion release from roots has not indicated a large role in P acquisition by these species and does not explain their contrasting critical P requirements (Kidd et al., 2016).

6.4.5 The role of arbuscular mycorrhizal fungi

Like many pasture legume species, the roots of *T. subterraneum* and *Ornithopus* spp. are usually found to be highly colonised by AMF (Smith et al., 2015; Sandral et al., 2019). Growth and P acquisition by *T. subterraneum* in P-deficient soil is usually improved by its association with AMF, but the benefit diminishes when soil P fertility and the P status of the host plant are improved (Abbott and Robson, 1977; Schweiger et al., 1995; Hill et al., 2010; McLachlan et al., 2020). In some cases, AMF colonisation of *T. subterraneum* has been associated with a lower critical P requirement relative to non-mycorrhizal plants (e.g. Schweiger et al., 1995), while in other cases the impact of AMF on critical P requirement has been marginal or non-existent, even when growth in P-deficient soil has clearly been improved (e.g. Abbot and Robson, 1977; Bolan et al., 1987; McLachlan et al., 2020).

Mycorrhizae colonise the roots of *Ornithopus* spp. at frequencies similar to those of *T. subterraneum* (e.g. Sandral et al., 2019). However, *Ornithopus* spp. appear to gain only a comparatively small P acquisition benefit from their AMF association (Schweiger et al., 1995; McLachlan et al., 2020). It has been hypothesised that the contrast in the AMF benefit obtained by these two legumes is due to the difference in their root hair lengths (i.e. the species with longer root hairs benefits less; Schweiger et al., 1995). However, it is difficult to reconcile this hypothesis with the observation that P acquisition by AMF occurs at much greater distances from the root surface than the length of root hairs. For example, most P acquisition by AMF occurs within ~1 cm of the root surface, with some P uptake also feasible at many centimetres distance (Jakobsen et al., 1992).

Likewise, the large difference in root hair length between *T. subterraneum* and *Ornithopus* spp. appears to largely explain their contrasting abilities to acquire P from low-P soil (Haling et al., 2016b). However, the ability of AMF to acquire P from soil at distances from the root that are a magnitude greater than the length of serradella root hairs (Jakobsen et al., 1992) should also negate the difference in P acquisition efficiency by these legumes. This does not occur. Subterranean clover exhibits significantly poorer P-acquisition efficiency and a significantly higher critical P requirement than serradella when colonised by AMF in the field (Sandral et al., 2019), and in glasshouse experiments in pasteurised media with and without AMF colonisation (Schweiger et al., 1995; McLachlan et al., 2020). The answer to this apparent anomaly lies, in part, in the fact that the clover reacts to AMF colonisation by allocating less root mass to nutrient foraging. This counteracts some of the anticipated P acquisition benefit of the AMF, and the differences in critical P requirements between *T. subterraneum* and serradella are preserved (McLachlan et al., 2020). It is notable that lower critical-P requirements detected for other alternative legumes in glasshouse experiments with low AMF colonisation rates, were not always detected consistently in the field (Fig. 7; Sandral et al., 2018; Sandral et al., 2019). This may be a consequence of AMF colonisation in the field, but is also likely to be a consequence of the many other challenges to root growth (soil hardness, soil acidity, dry topsoil, root diseases, etc.) in field environments (e.g. Watt et al., 2006; Simpson et al., 2011b). These challenges can reduce soil exploration and nutrient foraging by roots and may not be resisted or tolerated equally by different plant genotypes (e.g. Delhaize et al., 2009).

6.5 Is a region-wide shift to phosphorus-efficient pastures possible?

6.5.1 Requirements for soil phosphorus fertility management

As outlined already, soil test data are intrinsically variable due mainly to seasonal fluctuations in extractable P concentrations (Fig. 5). Farmers set a target range for their management of soil P that is guided by the critical P requirement of the grazing system. This typically has a peak-to-peak amplitude of about five Olsen P units (e.g. Fig. 5a; Simpson et al., 2009). The difference between the critical P requirements of *T. subterraneum* and the alternative P-efficient legumes identified by Sandral et al. (2019) was of similar size (Fig. 7) and, consequently, also suited the practical realities of managing soil fertility. The maximum benefit of using a P-efficient legume in a fertilised production system will only be realised if the target for soil P management is reduced to achieve the lower critical P requirement of the novel legume. It is questionable whether critical STP differences, less than five Olsen P units (even if statistically proven in field experiments), could be managed effectively enough on farms to deliver real savings in P-fertiliser costs or reduced environmental risks.

6.5.2 Edaphic requirements

Yellow serradella (*O. compressus*) was the first species of serradella to be commercially released in Australia (late 1950s). The early cultivars were intended for use on acid, infertile sandy soils where their deep roots, high tolerance of acid soils and hardseeds (transiently impermeable seed coats that protect the seeds against premature germination after summer storms) allowed them to be productive and persistent where *T. subterraneum* was failing (Bolland and Gladstones, 1987; Loi et al., 2005; Freebairn, 1996). Cultivars of French serradella (*O. sativus*) have been released for use since the mid 1990s also with the intent that they be used mainly on light textured soils. Serradella use was initially constrained by the relatively high cost of seed. However, the development of improved yellow and French serradella cultivars has underpinned their expanded use in rotation with crops since about 2009 (Nichols et al., 2012). A major breakthrough being the development of the first hardseeded cultivars of French serradella (Nutt, 2004a,b). This is an essential characteristic needed for persistence of the species in pastures that regularly experience drought (Dear et al., 2002). The expanding use of *Ornithopus* spp. in cropping systems demonstrated that the *Ornithopus* spp. were not 'niche' species only suited to sandy soils as thought originally. It is now recognised that they can be grown successfully on a much wider range of soil types where they can be highly productive, nodulate prolifically, and provide a forage with high feeding value that is potentially 'bloat-free' (Freebairn, 1996; Hackney et al., 2013).

6.5.3 Yield and persistence

In the experiments of Sandral et al. (2019) all of the P-efficient alternative legumes, except cultivars of *O. compressus*, produced peak spring yields that were equivalent to, or exceeded subterranean clover. However, *O. compressus* has been found to yield as well as *T. subterraneum* in other soil type/climatic districts (e.g. Bolland and Paynter, 1992). This highlights the continuing need to consider the fit of the P-efficient species to specific environments. In early experiments, the lower P-fertiliser requirement of *O. compressus* relative to *T. subterraneum* was also noted (e.g. Bolland, 1986b; Paynter, 1990; Bolland and Paynter, 1992). However, this did not lead to changed recommendations for soil P management because critical STP concentrations were not determined until the experiments of Sandral et al. (2019).

6.5.4 Further steps to achieving region-wide adoption of phosphorus-efficient serradella pastures

Phosphorus efficiency will only be captured by managing soil P fertility at the lower critical STP concentration suitable for *Ornithopus* spp. Research and extension to inform farm advisors and farmers who already grow *Ornithopus* spp. pastures that they can now adopt revised, species-specific STP benchmarks (Sandral et al., 2019) is progressing. However, expansion of *Ornithopus* spp. use into much larger areas of the permanent pasture zone where they have not been widely used, will be necessary to achieve a region-wide change in P use efficiency. This will depend on having cultivars that can deliver high yields and persistence across the diverse climatic zones of southern Australia. While there is evidence that some serradella cultivars can yield at least as well as *T. subterraneum* (Sandral et al., 2019), concerns that current cultivars may lack persistence in the new farming zones to which they are being targeted (e.g. Hayes et al., 2015) have not yet been addressed. Persistence of annual legumes is largely a function of their seed production, seed protection mechanisms, and seedling emergence and survival characteristics. Successful cultivars display stable flowering dates (irrespective of wide and unpredictable variations in the time of germination each year) that avoid frost damage at flowering and optimise seed production (Donald, 1970). Stable flowering dates among cultivars of *T. subterraneum* are regulated by their responses to long days (photoperiodism) and cold winter temperatures (vernalisation), and the optimum time for flowering is also well-defined (Aitken and Drake, 1941; Donald, 1970). Early research into these aspects of persistence among serradella cultivars indicates that some current cultivars may not display equivalent flowering date stability (Boschma et al., 2019). Seeds must also be protected against false breaks to the growing season by being initially impermeable to moisture (hardseeded), yet responsive to reliable rainfall when it occurs. Work on these cultivar traits has been in process for some time (e.g. Loi et al., 2005), but adaptations that are a better fit to new districts may also need to be pursued.

6.6 Summary: how research has contributed to the development of a more phosphorus-efficient and sustainable production system

In this case study, there were a number of essential steps necessary for the conception, development, and realisation of more P-efficient grassland farming.

6.6.1 Identifying opportunities to improve the efficiency of phosphorus fertiliser use

It is hard to know where to commence the discourse on how P efficiency has been improved in the grasslands of southern Australia because there has been a long history of incremental improvement in pasture production punctuated by step changes in progress as 'break-through' technologies emerged, all of which have ensured improving P use efficiency. Significant steps in this revolution have included the realisation that *T. subterraneum* was highly adapted to the southern Australian environment and could be widely adapted for use by developing cultivars to fit the rainfall gradient (i.e. length of growing season) that exists across the farming zone. Subsequently, there has been an effort to identify alternative legumes to diversify pasture composition and extend the N-fixing benefit of legumes to wider environments. Combining the use of P fertiliser and pasture legumes dramatically boosted P and N fertility across the region and delivered a step-change in production. Further innovations included recognition and alleviation of micronutrient deficiencies in soils, development of productive and persistent companion grasses, and the implementation of various fertiliser technologies to underpin more effective use of P on soils that are almost universally P-deficient and accumulate P when fertilised. Readers are referred to other sources for a more complete discussion of the revolution that has occurred over the last century (e.g. Rossiter, 1966; Moore, 1970b; Morley, 1961; Costin and Williams, 1983; Smith, 2000; Nichols et al., 2007, 2012; McLaughlin et al., 2011).

While the initial emphasis of research was to correct soil-nutrient deficiencies to enable improved pasture production and livestock carrying capacity, the emphasis in the modern era has shifted to the optimal use of P, guided by the critical P requirement of the soil-pasture-livestock system (e.g. Figs 2 and 5; Simpson et al., 2009; Gourley et al., 2019). Removing soil fertility limitations ensures the optimal use of Australia's most-limiting resources: water and land (e.g. Mills et al., 2006) and underpins the profitability and sustainability of farm businesses (Lean et al., 1997). Over-use of P fertiliser still occurs in some sectors, but it is actively discouraged (e.g. Gourley and Weaver, 2012).

Optimal management of P inputs to pastures growing on P-sorbing soils inevitably required P inputs that are greater than the amount of P that is exported in agricultural products. In many ways this is a 'technical' inefficiency that has seemed to be unavoidable because of the moderate to high P-sorbing nature of the most Australian soils. However, inefficiency creates opportunity. Steady improvement in the understanding of how and why P accumulates in fertilised soils, and the form and lability of accumulated P (e.g. Barrow, 1974, 1999; Simpson et al., 2015; Jarosch et al., 2015; McLaren et al., 2017b; Barrow et al., 2018) has opened a new opportunity to manage P accumulations in soil by using plants with roots that acquire P more effectively and support a lower critical P requirement for pasture production (e.g. Sandral et al., 2019).

6.6.2 Development of knowledge and tools to enable application of the science

Pastures that have a lower critical P requirement only deliver a reduced P-fertiliser requirement when the soils in which they are growing are fertilised to the lower critical STP concentration of the novel pasture. The saving in P-fertiliser is realised indirectly because at the lower STP concentration, P accumulation reactions in soil are slower and the risk of P losses via runoff and leaching is reduced. Consequently, it was crucial that farmers were equipped with the knowledge and tools to be able to fertilise soils to meet critical STP targets (e.g. Fig. 5). This is now promoted as best practice in the target area of this case study (e.g. Simpson et al., 2009). However, it is sobering to remember the timeframe that was required for research, development and adoption of this essential step. Soil tests were developed in the 1950–1960s (Olsen et al., 1954; Colwell, 1963), development and adoption of soil testing was an incremental process (e.g. Reuter et al., 1995), and their application on farms Australia-wide was finally enabled in 2007 with the release of reliable, widely applicable STP benchmarks (Gourley et al., 2019).

6.6.3 Identifying legumes that can underpin a step-change in P efficiency of the farming system

A search for alternative, high-yielding pasture legumes with substantially lower critical STP requirements than that of *T. subterraneum* was commenced soon after their potential P-efficiency advantages were realised (Simpson et al., 2014). Success in this task was only possible, however, because alternative pasture legumes had already been trialled over many years in southern Australia with the objective of diversifying the feed base for animal production (Nichols et al., 2007; Nichols et al., 2012). Despite a large number of potential candidates, the number of legumes that proved to be consistently P-efficient under field conditions was limited (Sandral et al., 2019). Only two P-efficient species potentially suited to use in permanent pastures were identified (*O. compressus* and *O. sativus*). Some of the barriers to their adoption were readily crossed because they were already in limited use with acknowledged agronomic and animal feed value (Nichols et al., 2007; Nichols et al., 2012). Widespread adoption will be essential for a real step-change in P efficiency of the grassland system. This now depends on the identification or development of cultivars that are adapted to the newer soil and climatic environments in which they will need to be grown and assurance that they can meet modern, socially fit-for-purpose criteria (e.g. nutritious, food-chain safe, low weed risk, etc.; Revell and Revell, 2007).

7 Conclusion and future trends in research

The pursuit of high PBE by improving nutrient foraging has natural limitations. For example, examination of a wide range of *T. subterraneum* genotypes and among phylogenetically allied *Trifolium* spp. (McLachlan et al., 2019; McLachlan et al., 2020; and authors' unpublished data) has failed, so far, to find variation in P-efficient root morphology traits (especially root hair length) that are equivalent to those of the serradella species. There are natural limits to the range in phenes within a species that reflect the environments in which they have evolved. In addition to this, there are practical limits imposed by the agricultural system being improved. In the mixed (legume-grass-forb) grasslands of southern Australia, the critical P requirement for production is effectively determined by the pasture component with the highest P requirement (currently the legume component; Hill et al., 2010). If widespread replacement of the dominant legume (*T. subterraneum*) with *Onithopus* spp. can be achieved (as discussed above), the critical P requirement of the grassland system will be reduced and will be roughly similar to that of key companion grasses (e.g. *Phalaris aquatica* L., *Dactylis glomerata* L.; Sandral et al., 2019). Although there are several other grassland component species with even lower critical P requirements (Hill et al., 2010), pursuing further improvement in legume P-acquisition efficiency beyond that delivered by *Ornithopus* spp. would be naive and futile because such a change would also demand new varieties of *P. aquatica* and *D. glomerata* with equally improved P efficiency. This is impractical for the foreseeable future.

Mixed grasslands are natural intercrops which already thrive through complementary access to N-fixation delivered via their legume component. Consequently, an alternative approach to further reduce the critical P requirement of the 'pasture system' as a whole, may be to introduce species that utilise a 'P-mining' acquisition strategy with the aim of capturing complementary and/or facilitated P acquisition for grassland production (e.g. analogous to the *Vicia faba-Zea mays* intercrop system; Li et al., 2003). Exploratory experiments using *Lupinus albus* as a test species suggest that mixing plants with 'P-mining' and 'P-foraging' strategies may lower the critical P requirement of a pasture, but a field experiment, whilst promising, was not conclusive (e.g. Becquer et al., 2017; Sandral et al., 2019). There is also evidence from numerous experiments of shared P access among interplanted crop species (e.g. Li et al., 2014b; Gardner and Boundy, 1983; Horst et al., 2001; Cu et al., 2005), but further work in pastures is limited by a lack of conventional pasture species with unequivocal P-mining attributes, and generally because of inconclusive or negative attempts to improve P acquisition of crop species *in soil* by increasing organic acid anion or phosphatase secretion from roots (e.g. for further discussion, see Richardson et al., 2011; Ryan et al., 2014).

Achievement of unequivocal 'P-mining' attributes in conventional crop and pasture varieties and even the integration of P-mining benefits into current agricultural systems (using species that include *L. albus*, etc.; e.g. Mat Hassan et al., 2013; Doolette et al., 2019) remain as major research challenges for improved P-acquisition efficiency in agriculture.

8 Where to look for further information

For classic texts on P supply and P acquisition by plants:

- Barber, S. A. (1995), *Soil nutrient bioavailability; a mechanistic approach*. John Wiley and Sons, New York, USA.
- Tinker, P. B. and Nye, P. H. (2000), *Solute movement in the rhizosphere*. Oxford University Press Inc., New York, USA.

For a broader understanding of the global P cycle and the impacts of anthropogenic activity:

- Filippelli, G. M. (2008), The global phosphorus cycle: past, present, and future. *Elements* 4, 89-95.
- Föllmi, K. B. (1996), The phosphorus cycle, phosphogenesis and marine phosphate-rich deposits. *Earth-Science Reviews* 40, 55-124.

For practical guides on the management of P fertiliser and the use of critical soil test values:

- Gourley, C. J. P., Melland, A. R., Waller, R. A., Awty, I. M., Smith, A. P., Peverill, K. I. and Hannah, M. C. (2007), *Making better fertiliser decisions for grazed pastures in Australia*. Victorian Government Department of Primary Industries, Melbourne. Currently available on the CSIRO Australian Soil Resource Information System website at: https://www.asris.csiro.au/dow nloads/BFD/Making%20Better%20Fertiliser%20Decisions%20for%20 Grazed%20Pastures%20in%20Australia.pdf.

  Results supporting this resource are found in: Gourley, C. J. P., Weaver, D. M., Simpson, R. J., Aarons, S. R., Hannah, M. M. and Peverill, K. I. (2019), The development and application of functions describing pasture yield responses to phosphorus, potassium, and sulfur in Australia using meta-data analysis and derived soil-test calibration relationships. *Crop Pasture Sci.* 70, 1065-1079.
- *Making Better Fertiliser Decisions for Cropping Systems in Australia*. An online resource and tool including specific case studies for phosphorus. Currently available: www.bfdc.com.au. Associated material is available on

the Grains Research and Development Corporation website at: www.grdc
.com.au (e.g. November 2013 Crop Nutrition Fact Sheet "Northern, Southern
and Western Regions: Better Fertiliser Decisions for Crop Nutrition";
https://grdc.com.au/__data/assets/pdf_file/0025/194443/grdc_
fs_better-fertiliser-decisions-for-crop-nutrition_high-res-pdf.pdf.pdf).
- Simpson, R., Graham, P., Davies, L., Zurcher, E. (2009). *Five easy steps to ensure you are making money from superphosphate*. Decision support tool. CSIRO & Industry and Investment NSW: Sydney. Currently available on the Meat and Livestock Australia website at: https://www.mla.com.au/globalassets/mla-corporate/generic/extension-training-and-tools/5-easy-steps-guide.pdf.

For a useful general discussion of the principles of P use efficiency in agriculture and the principle of the Four Rs of nutrient stewardship:

- Roberts, T. L., Johnston, A. E. (2015), Phosphorus use efficiency and management in agriculture. *Resour. Conserv. Recycl.* 105, 275-281.
  (*Note: the definition of critical P requirement used by Roberts and Johnston (2015) is different to that used in the current chapter and can only be achieved in soils with low or negligible P-sorption capacities*).
- *The Four Rs of nutrient stewardship* (developed collaboratively by The Fertilizer Institute, International Plant Nutrition Institute, Fertilizer Canada, International Fertilizer Association). Information currently available at: http://www.ipni.net/4R.

For a comprehensive overview of the pasture systems of Australia, including those described in the case study of this chapter:

- Moore, R. M. (1970), Australian Grasslands. Australian National University Press: Canberra, ACT.

For information regarding improvements to the internal P use efficiency of plants, a topic not covered by the current chapter:

- Veneklaas, E. J., Lambers, H., Bragg, J., Finnegan, P. M., Lovelock, C. E., Plaxton, W. C., Price, C. A., Scheible, W. R., Shane, M. W., White, P. J. and Raven, J. A. (2012), Opportunities for improving phosphorus-use efficiency in crop plants. *New Phytol.* 195, 306-320.
- Rose, T. J., Liu, L. and Wissuwa, M. (2013), Improving phosphorus efficiency in cereal crops: Is breeding for reduced grain phosphorus concentration part of the solution? *Front. Plant. Sci.* 4, 444.
- Rose, T. J. and Wissuwa, M. (2012), Rethinking internal phosphorus utilization efficiency: a new approach is needed to improve PUE in grain crops. *Adv. Agron.* 116, 185-217.

9 Acknowledgements

RJS and REH are members of a research team examining the development of 'Phosphorus Efficient Pastures' with funding from the Australian Government Department of Agriculture, Water and the Environment as part of its Rural R&D for Profit program, Meat and Livestock Australia, Dairy Australia, Australian Wool Innovations Ltd, and the participating research organisations and farmer groups.

10 References

Abbott, L. K. and Robson, A. D. (1977). Growth stimulation of subterranean clover with vesicular arbuscular mycorrhizas. *Aust. J. Agric. Res.* 28(4), 639-649.

Aitken, Y. and Drake, F. R. (1941). Studies on the varieties of subterranean clover. *Proc. R. Soc. Victoria* 53, 342-393.

Bai, Z., Li, H., Yang, X., Zhou, B., Shi, X., Wang, B., Li, D., Shen, J., Chen, Q., Qin, W., Oenema, O. and Zhang, F. (2013). The critical soil P levels for crop yield, soil fertility and environmental safety in different soil types. *Plant Soil* 372(1-2), 27-37.

Barber, S. A. (1995). *Soil Nutrient Bioavailability: A Mechanistic Approach*. New York: John Wiley and Sons.

Barrow, N. J. (1974). The slow reactions between soil and anions. I. Effects of soil, temperature and water content of a soil on the decrease in effectiveness of phosphate for plant growth. *Soil Sci.* 118(6), 380-386.

Barrow, N. J. (1980a). Differences among some North American soils in the rate of reaction with phosphate. *J. Environ. Qual.* 9(4), 644-648.

Barrow, N. J. (1980b). Differences amongst a wide-ranging collection of soils in the rate of reaction with phosphate. *Aust. J. Soil. Res.* 8, 215-224.

Barrow, N. J. (1999). The four laws of soil chemistry: the Leeper lecture 1998. *Soil Res.* 37(5), 787-829.

Barrow, N. J. (2015). Soil phosphate chemistry and the P-sparing effect of previous phosphate applications. *Plant Soil* 397(1-2), 401-409.

Barrow, N. J., Bolland, M. D. A. and Allen, D. G. (1998). Effect of previous additions of superphosphate on sorption of phosphate. *Soil Res.* 36(3), 359-372.

Barrow, N. J. and Debnath, A. (2014). Effect of phosphate status on the sorption and desorption properties of some soils of northern India. *Plant Soil* 378(1-2), 383-395.

Barrow, N. J., Debnath, P. B. A. and Debnath, A. (2018). Three residual benefits of applying phosphate fertilizer. *Soil Sci. Soc. Am. J.* 82(5), 1168-1176.

Bates, T. R. and Lynch, J. P. (1996). Stimulation of root hair elongation in *Arabidopsis thaliana* by low phosphorus availability. *Plant Cell Environ.* 19(5), 529-538.

Becquer, A., Haling, R. E., Stefanski, A., Richardson, A. E. and Simpson, R. J. (2017). Complementary phosphorus acquisition strategies of interplanted subterranean clover and white lupin increase sward yield in a low phosphorus soil. *Proceedings of the 18th Aust. Soc. Agron. Conference*, Ballarat, Australia. Available at: http://agronomyaustraliaproceedings.org/images/sampledata/2017/199_ASA2017_Becquer_Adeline_Final.pdf.

Bell, L. W., Ryan, M. H., Bennett, R. G., Collins, M. T. and Clarke, H. J. (2012). Growth, yield and seed composition of native Australian legumes with potential as grain crops. *J. Sci. Food Agric.* 92(7), 1354-1361.

Bell, M. J., Moody, P. W., Anderson, G. C. and Strong, W. (2013). Soil phosphorus - crop response calibration relationships and criteria for oilseeds, grain legumes and summer cereal crops grown in Australia. *Crop Pasture Sci.* 64(5), 499–513.

Benbi, D. K. and Gilkes, R. J. (1987). The movement into soil of P from superphosphate grains and its available to plants. *Fert. Res.* 12(1), 21–36.

Blackwell, M. S. A., Brookes, P. C., de la Fuente-Martinez, N., Murray, P. J., Snars, K. E., Williams, J. K. and Haygarth, P. M. (2009). Effects of soil drying and rate of re-wetting on concentrations and forms of phosphorus in leachate. *Biol. Fertil. Soils* 45(6), 635–643.

Bolan, N. S., Robson, A. D. and Barrow, N. J. (1987). Effects of vesicular-arbuscular mycorrhiza on the availability of iron phosphates to plants. *Plant Soil* 99(2–3), 401–410.

Bolland, M. D. A. (1986a). Residual value of phosphorus from superphosphate for wheat grown on soils of contrasting texture near Esperance, Western Australia. *Aust. J. Exp. Agric.* 26, 209–215.

Bolland, M. D. A. (1986b). Efficiency with which yellow serradella and subterranean clover use superphosphate on a deep sandy soil near Esperance, Western Australia. *Aust. J. Exp. Agric.* 26, 675–679.

Bolland, M. D. A. and Baker, M. J. (1998). Phosphate applied to soil increases the effectiveness of subsequent applications of phosphate for growing wheat shoots. *Aust. J. Exp. Agric.* 38(8), 865–869.

Bolland, M. D. A. and Gladstones, J. S. (1987). Serradella (*Ornithopus* spp.) as a pasture legume in Australia. *J. Aust. Inst. Agric. Sci.* 53, 5–10.

Bolland, M. D. A. and Paynter, B. H. (1992). Comparative responses of annual pasture legume species to superphosphate applications in medium and high rainfall areas of Western Australia. *Fert. Res.* 31(1), 21–33.

Borkert, C. M. and Barber, S. A. (1985). Predicting the most efficient phosphorus placement for soybeans. *Soil Sci. Soc. Am. J.* 49(4), 901–904.

Boschma, S., Kidd, D., Newell, M., Stefanski, A., Haling, R., Hayes, R., Ryan, M. and Simpson, R. (2019). Flowering time responses of serradella cultivars. *Proceedings of the 19th Aust. Soc. Agron. Conference*, Wagga Wagga, Australia. Available at: http://agr onomyaustraliaproceedings.org/images/sampledata/2019/2019ASA_Simpson_R ichard_169.pdf.

Bouwman, L., Goldewijk, K. K., Van Der Hoek, K. W., Beusen, A. H. W., Van Vuuren, D. P., Willems, J., Rufinoe, M. C. and Stehfesta, E. (2013). Exploring global changes in nitrogen and phosphorus cycles in agriculture induced by livestock production over the 1900-2050 period. *Proc. Natl Acad. Sci. U.S.A.* 110(52), 20882–20887.

Bowen, G. D. and Rovira, A. D. (1999). The rhizosphere and its management to improve plant growth. *Adv. Agron.* 66, 1–102.

Brouwer, R. (1962). Nutritive influences on the distribution of dry matter in the plant. *Neth. J. Agric. Sci.* 10(5), 399–408.

Burkitt, L. L., Moody, P. W., Gourley, C. J. P. and Hannah, M. C. (2002). A simple phosphorus buffering index for Australian soils. *Aust. J. Soil Res.* 40, 1–18.

Burkitt, L. L., Sale, P. W. G. and Gourley, C. J. P. (2008). Soil phosphorus buffering measures should not be adjusted for current phosphorus fertility. *Aust. J. Soil Res.* 46(8), 676–685.

Burridge, J. D., Findeis, J. L., Jochua, C. N., Miguel, M. A., Mubichi-Kut, F. M., Quinhentos, M. L., Xerinda, S. A. and Lynch, J. P. (2019). A case study on the efficacy of root phenotypic selection for edaphic stress tolerance in low-input agriculture: common bean breeding in Mozambique. *Field Crops Res.* 244, 107512. doi:10.1016/j. fcr.2019.107612.

Campbell, B. M., Beare, D. J., Bennett, E. M., Hall-Spencer, J. M., Ingram, J. S. I., Jaramillo, F., Ortiz, R., Ramankutty, N., Sayer, J. A. and Shindell, D. (2017). Agriculture production as a major driver of the Earth system exceeding planetary boundaries. *Ecol. Soc.* 22(4), 8. doi:10.5751/ES-09595-220408.

Carter, E. D. and Day, H. R. (1970). Interrelationships of stocking rate and superphosphate rate on pasture as determinants of animal production. I. Continuously grazed old pasture land. *Aust. J. Agric. Res.* 21(3), 473–491.

Cayley, J. W. D., Kearney, G. A., Saul, G. R. and Lescun, C. L. (1999). The long-term influence of superphosphate and stocking rate on the production of spring-lambing Merino sheep in the high rainfall zone of southern Australia. *Aust. J. Agric. Res.* 50(7), 1179–1190.

Chen, C. R., Condron, L. M., Davis, M. R. and Sherlock, R. R. (2002). Phosphorus dynamics in the rhizosphere of perennial ryegrass (*Lolium perenne* L.) and radiata pine (*Pinus radiata* D.Don). *Soil Biol. Biochem.* 34(4), 487–499.

Colwell, J. D. (1963). The estimation of the phosphorus fertilizer requirements of wheat in southern New South Wales by soil analysis. *Aust. J. Exp. Agric.* 3(10), 190–197.

Cordell, D., Drangert, J.-O. and White, S. (2009). The story of phosphorus: global food security and food for thought. *Glob. Environ. Chang.* 19(2), 292–305.

Cornish, P. S. and Myers, L. F. (1977). Low pasture productivity of a sedimentary soil in relation to phosphate and water supply. *Aust. J. Exp. Agric.* 17(88), 776–783.

Costin, A. B. and Williams, C. H. (1983). *Phosphorus in Australia*. Canberra, Australia: Australian National University, Centre for Resource and Environmental Studies.

Cruz-Paredes, C., Svenningsen, N. B., Nybroe, O., Kjoller, R., Froslev, T. G. and Jakobsen, I. (2019). Suppression of arbuscular mycorrhizal fungal activity in a diverse collection of non-cultivated soils. *FEMS Microbiol. Ecol.* 95(3), 1–10.

Cu, S. T. T., Hutson, J. and Schuller, K. A. (2005). Mixed culture of wheat (*Triticum aestivum* L.) with white lupin (*Lupinus albus* L.) improves the growth and phosphorus nutrition of the wheat. *Plant Soil* 272(1-2), 143–151.

Cullen, B. R., Eckard, R. J., Callow, M. N., Johnson, I. R., Chapman, D. F., Rawnsley, R. P., Garcia, S. C., White, T. and Snow, V. O. (2008). Simulating pasture growth rates in Australian and New Zealand grazing systems. *Aust. J. Agric. Res.* 59(8), 761–768.

Culvenor, R. A. and Simpson, R. J. (2016). Interaction of plant genotype and management in the persistence of a perennial grass exposed to grazing and soil fertility stresses. *Grass Forage Sci.* 71(4), 540–558.

Curll, M. L. (1977). Superphosphate on perennial pastures. I. Effect of a pasture response on sheep production. *Aust. J. Agric. Res.* 28(6), 991–1005.

Dear, B. S., Wilson, B. C. D., Rodham, C. A., McCaskie, P. and Sandral, G. A. (2002). Productivity and persistence of *Trifolium hirtum, T. michelianum, T. glanduliferum* and *Ornithopus sativus* sown as monoculture or in mixtures with *T. subterraneum* in the south-eastern Australian wheat belt. *Aust. J. Exp. Agric.* 42(5), 549–556.

Delhaize, E., Taylor, P., Hocking, P. J., Simpson, R. J., Ryan, P. R. and Richardson, A. E. (2009). Transgenic barley (*Hordeum vulgare* L.) that express the wheat aluminium resistance gene (*TaALMT1*) show enhanced phosphorus nutrition and grain production when grown on an acid soil. *Plant Biotechnol. J.* 7(5), 391–400.

Denton, M. D., Sasse, C., Tibbett, M. and Ryan, M. H. (2006). Root distributions of Australian herbaceous perennial legumes in response to phosphorus placement. *Funct. Plant Biol.* 33(12), 1091–1102.

Dessureault-Rompre, J., Nowack, B., Schulin, R. and Luster, J. (2007). Spatial and temporal variation in organic acid anion exudation and nutrient anion uptake in the rhizosphere of *Lupinus albus* L. *Plant Soil* 301(1–2), 123–134.

Dibb, D. W., Fixen, P. E. and Murphy, L. S. (1990). Balanced fertilization with particular reference to phosphates: interaction of phosphorus with other inputs and management practices. *Fert. Res.* 26(1–3), 29–52.

Dinkelaker, B., Romheld, B. V. and Marschner, H. (1989). Citric acid excretion and precipitation of Ca citrate in the rhizosphere of white lupin (*Lupinus albus* L.). *Plant Cell. Environ.* 12, 285–292.

Dodd, J. R. and Mallarino, A. P. (2005). Soil-test phosphorus and crop grain yield responses to long-term phosphorus fertilisation for corn-soybean rotations. *Soil Sci. Soc. Am. J.* 69(4), 1118–1128.

Donald, C. M. (1970). Temperate pasture species. In: Moore, R. M. (Ed.), *Australian Grasslands*. Canberra, Australia: Australian National University Press, pp. 303–320.

Donald, C. M. and Williams, C. H. (1954). Fertility and productivity of a podzolic soil as influenced by subterranean clover (*Trifolium subterraneum* L.) and superphosphate. *Aust. J. Agric. Res.* 5(4), 664–687.

Doolette, A., Armstrong, R., Tang, C., Guppy, C., Mason, S. and McNeill, A. (2019). Phosphorus uptake benefit for wheat following legume break crops in semi-arid Australian farming systems. *Nutr. Cycl. Agroecosyst.* 113(3), 247–266.

Doyle, P. T., Grimm, M. and Thompson, A. N. (1993). Grazing for pasture and sheep management in the annual pasture zone. In: Kemp, D. R. and Michalk, D. L. (Eds) *Pasture Management Technology for the 21st Century*. Melbourne: CSIRO Australia, pp. 71–90.

Dunbabin, V., Diggle, A. and Rengel, Z. (2003). Is there an optimal root architecture for nitrate capture in leaching environments? *Plant Cell Environ.* 26(6), 835–844.

Duncan, W. G. and Ohlrogge, A. J. (1958). Principles of nutrient uptake from fertilizer bands. II. Root development in the band. *Agron. J.* 50, 605–608.

Eissenstat, D. M. (1991). On the relationship between specific root length and rate of root proliferation: a field study using citrus rootstocks. *New Phytol.* 118, 63–68.

Erisman, J. W., Sutton, M. A., Galloway, J. N., Klimont, Z. and Winiwarter, W. (2008). How a century of ammonia synthesis changed the world. *Nat. Geosci.* 1(10), 636–639.

Filippelli, G. M. (2008). The global phosphorus cycle: past, present, and future. *Elements* 4(2), 89–95.

Fischer, R. A. and Connor, D. J. (2018). Issues for cropping and agricultural science in the next 20 years. *Field Crops Res.* 222, 121–142.

Föhse, D. and Jungk, A. (1983). Influence of phosphate and nitrate supply on root hair formation of rape, spinach and tomato plants. *Plant Soil* 74(3), 359–368.

Föllmi, K. B. (1996). The phosphorus cycle, phosphogenesis and marine phosphate-rich deposits. *Earth Sci. Rev.* 40(1–2), 55–124.

Freebairn, R. D. (1996). *The History of Serradella (Ornithopus spp.) in NSW – A Miracle Pasture for Light Soils. Agdex 137*. Sydney: NSW Agriculture and Fisheries.

Freschet, G. T., Swart, E. M. and Cornelissen, J. H. C. (2015). Integrated plant phenotypic responses to contrasting above- and below-ground resources: key roles of specific leaf area and root mass fraction. *New Phytol.* 206(4), 1247–1260.

Gahoonia, T. S. and Nielsen, N. E. (1992). The effect of root induced pH changes on the depletion of inorganic and organic phosphorus in the rhizosphere. *Plant Soil* 143(2), 185–191.

Gahoonia, T. S., Care, D. and Nielsen, N. E. (1997). Root hairs and phosphorus acquisition of wheat and barley cultivars. *Plant Soil* 191(2), 181–188.

Gamuyao, R., Chin, J. H., Pariasca-Tanaka, J., Pesaresi, P., Dalid, C., Slamet-Loedin, I., Tecson-Mendoza, E. M., Wissuwa, M. and Heuer, S. (2012). The protein kinase OsPSTOL1 from traditional rice confers tolerance of phosphorus deficiency. *Nature* 488, 535–539.

Gardner, W. K. and Boundy, K. A. (1983). The acquisition of phosphorus by *Lupinus albus* L. IV. The effect of interplanting wheat and white lupin on the growth and mineral composition of the two species. *Plant Soil* 70(3), 391–402.

George, T. S., Gregory, P. J., Robinson, J. S., Buresh, R. J. and Jama, B. (2002). Utilisation of soil organic P by agroforestry and crop species in the field, western Kenya. *Plant Soil* 246(1), 53–63.

George, T. S., Hocking, P. J., Gregory, P. J. and Richardson, A. E. (2008). Variation of root-associated phosphatase in wheat cultivars explains their ability to utilise organic P substrates in-vitro, but does not effectively predict P-nutrition in a range soils. *Environ. Exp. Bot.* 64, 239–249.

Gerke, J. (1994). Kinetics of soil phosphate desorption as affected by citric acid. *Z. Pflanzenernaehr. Bodenk.* 157(1), 17–22.

Gourley, C. J. P., Aarons, S. R., Hannah, M. C., Awty, I. M., Dougherty, W. J. and Burkitt, L. L. (2015). Soil phosphorus, potassium and sulphur excesses, regularities and heterogeneity in grazing-based dairy farms. *Agric. Ecosyst. Environ.* 201, 70–82.

Gourley, C. J. P., Melland, A. R., Waller, R. A., Awty, I. M., Smith, A. P., Peverill, K. I. and Hannah, M. C. (2007). Making better fertiliser decisions for grazed pastures in Australia. Melbourne: Victorian Government Department of Primary Industries. Available at: http://www.asris.csiro.au/downloads/BFD/Making%20Better%20Fertili ser%20Decisions%20for%20Grazed%20Pastures%20in%20Australia.pdf.

Gourley, C. J. P. and Weaver, D. M. (2012). Nutrient surpluses in Australian grazing systems: management practices, policy approaches, and difficult choices to improve water quality. *Crop Pasture Sci.* 63(9), 805–818.

Gourley, C. J. P., Weaver, D. M., Simpson, R. J., Aarons, S. R., Hannah, M. M. and Peverill, K. I. (2019). The development and application of functions describing pasture yield responses to phosphorus, potassium and sulfur in Australia using meta-data analysis and derived soil-test calibration relationships. *Crop Pasture Sci.* 70(12), 1065–1079.

Graham, R. P. (2006). Bookham grazing demonstration results. *NSW DPI Sheep Conference.* Orange: New South Wales Department of Primary Industries, pp. 211–216.

Graham, P. (2017). Bookham grazing demonstration results. New South Wales: Department of Primary Industries. Available at: http://www.fscl.org.au/wp-content/u ploads/2017/09/Bookham-Grazing-demo-results-P-Graham.pdf.

Greenwood, K. L. and Hutchinson, K. J. (1998). Root characteristics of temperate pasture in New South Wales after grazing at three stocking rates for 30 years. *Grass Forage Sci.* 53(2), 120–128.

Hackney, B., Rodham, C. and Piltz, J. (2013). *Using French Serradella to Increase Crop and Livestock Production.* Sydney: Meat and Livestock Australia Ltd.

Haling, R. E., Brown, L. K., Stefanski, A., Kidd, D. R., Ryan, M. H., Sandral, G. A., George, T. S., Lambers, H. and Simpson, R. J. (2018). Differences in nutrient foraging among *Trifolium subterraneum* cultivars deliver improved P-acquisition efficiency. *Plant Soil* 424(1-2), 539-554.

Haling, R. E., Richardson, A. E., Culvenor, R. A., Lambers, H. and Simpson, R. J. (2010a). Root morphology, root-hair development and rhizosheath formation on perennial grass seedlings is influenced by soil acidity. *Plant Soil* 335(1-2), 457-468.

Haling, R. E., Simpson, R. J., Delhaize, E., Hocking, P. J. and Richardson, A. E. (2010b). Effect of lime on root growth, morphology and the rhizosheath of cereal seedlings growing in an acid soil. *Plant Soil* 327(1-2), 199-212.

Haling, R. E., Yang, Z. J., Shadwell, N., Culvenor, R. A., Stefanski, A., Ryan, M. H., Sandral, G. A., Kidd, D. R., Lambers, H. and Simpson, R. J. (2016a). Growth and root dry matter allocation by pasture legumes and a grass with contrasting external critical phosphorus requirements. *Plant Soil* 407(1-2), 67-79.

Haling, R. E., Yang, Z. J., Shadwell, N., Culvenor, R. A., Stefanski, A., Ryan, M. H., Sandral, G. A., Kidd, D. R., Lambers, H. and Simpson, R. J. (2016b). Root morphological traits that determine phosphorus-acquisition efficiency and critical external phosphorus requirement in pasture species. *Funct. Plant Biol.* 43(9), 815-826.

Handreck, K. A. (1997). Phosphorus requirements of Australian native plants. *Aust. J. Soil Res.* 35(2), 241-289.

Hayes, J. E., Simpson, R. J. and Richardson, A. E. (2000). The growth and utilisation of plants in sterile media when supplied with inositol hexaphosphate, glucose 1-phosphate or inorganic phosphate. *Plant Soil* 220(1/2), 165-174.

Hayes, R. C., Sandral, G. A., Simpson, R., Price, A., Stefanski, A. and Newell, M. T. (2015). A preliminary evaluation of alternative annual legume species under grazing on the Southern Tablelands of NSW. *Proceedings of the 17th Aust. Soc. Agron. Conference, Hobart*, Australia. Available at: http://agronomyaustraliaproceedings.org/images/sampledata/2015_Conference/pdf/agronomy2015final00332.pdf.

Hedley, M. and McLaughlin, M. (2005). Reactions of phosphate fertilizers and by-products in soils. In: Sims, J. T. and Sharpley, A. N. (Eds), *Phosphorus: Agriculture and the Environment, Agronomy Monograph No. 46*. Madison, WI: American Society of Agronomy, Crop Science Society of America, Soil Science Society of America, pp. 181-252.

Hedley, M. J., White, R. E. and Nye, P. H. (1982). Plant induced changes in the rhizosphere of rape (*Brassica napus* var. Emerald) seedlings. III Changes in L value, soil phosphate fractions and phosphatase activity. *New Phytol.* 91, 45-56.

Helyar, K. R. and Anderson, A. J. (1970). Responses of five pasture species to phosphorus, lime, and nitrogen on an infertile acid soil with a high phosphate sorption capacity. *Aust. J. Agric. Res.* 21(5), 677-692.

Helyar, K. and Spencer, K. (1977). Sodium bicarbonate soil test values and the phosphate buffering capacity of soils. *Aust. J. Soil Res.* 15(3), 263-273.

Helyar, K. R. and Price, G. H. (1999). Making recommendations based on soil tests. In: Peverill, K. I., Sparrow, L. A. and Reuter, D. J. (Eds), *Soil Analysis: An Interpretation Manual*. Collingwood, Australia: CSIRO Publishing.

Heuer, S., Gaxiola, R., Schilling, R., Herrera-Estrella, L., Lopez-Arredondo, D., Wissuwa, M., Delhaize, E. and Rouached, H. (2017). Improving phosphorus use efficiency: a complex trait with emerging opportunities. *Plant J.* 90(5), 868-885.

Hill, J. O., Simpson, R. J., Ryan, M. H. and Chapman, D. F. (2010). Root hair morphology and mycorrhizal colonisation of pasture species in response to phosphorus and nitrogen nutrition. *Crop Pasture Sci.* 61(2), 122-131.

Hill, J. O., Simpson, R. J., Wood, J. T., Moore, A. D. and Chapman, D. F. (2005). The phosphorus and nitrogen requirements of temperate pasture species and their influence on grassland botanical composition. *Aust. J. Agric. Res.* 56(10), 1027-1039.

Hill, M. J. and Donald, G. E. (1998). *Australian Temperate Pastures Database.* Perth, Western Australia: National Pasture Improvement Coordinating Committee/CSIRO Division of Animal Production (CD-ROM).

Hinsinger, P., Betencourt, E., Bernard, L., Brauman, A., Plassard, C., Shen, J., Tang, X. and Zhang, F. (2011). P for two, sharing a scarce resource: soil phosphorus acquisition in the rhizosphere of intercropped species. *Plant Physiol.* 156(3), 1078-1086.

Ho, M. D., Rosas, J. C., Brown, K. M. and Lynch, J. P. (2005). Root architectural tradeoffs for water and phosphorus acquisition. *Funct. Plant Biol.* 32(8), 737-748.

Holford, I. and Crocker, G. (1988). Efficacy of various soil phosphate tests for predicting phosphate responsiveness and requirements of clover pastures on acidic tableland soils. *Aust. J. Soil Res.* 26(3), 479-488.

Horst, W. J., Kamh, M., Jibrin, J. M. and Chude, V. O. (2001). Agronomic measures for increasing P availability to crops. *Plant Soil* 237(2), 211-223.

Hufnagel, B., de Sousa, S. M., Assis, L., Guimaraes, C. T., Leiser, W., Azevedo, G. C., Negri, B., Larson, B. G., Shaff, J. E., Pastina, M. M., Barros, B. A., Weltzien, E., Rattunde, H. F., Viana, J. H., Clark, R. T., Falcão, A., Gazaffi, R., Garcia, A. A., Schaffert, R. E., Kochian, L. V. and Magalhaes, J. V. (2014). Duplicate and conquer: multiple homologs of PHOSPHORUS-STARVATION TOLERANCE1 enhance phosphorus acquisition and sorghum performance on low phosphorus soils. *Plant Physiol.* 166(2), 659-677.

Jackson, T. and Malcolm, B. (2018). Using returns, risks and learning to understand innovation adoption. *Aust. Agribusiness Rev.* 26, 32-47.

Jakobsen, I., Abbott, L. K. and Robson, A. D. (1992). External hyphae of vesicular-arbuscular mycorrhizal fungi associated with *Trifolium subterraneum* L. 2. Hyphal transport of ^{32}P over defined distances. *New Phytol.* 120(4), 509-516.

Jarosch, K. A., Doolette, A. L., Smernik, R. J., Tamburini, F., Frossard, E. and Bünemann, E. K. (2015). Characterisation of soil organic phosphorus in NaOH-EDTA extracts: a comparison of ^{31}P NMR spectroscopy and enzyme addition assays. *Soil Biol. Biochem.* 91, 298-309.

Jarosch, K. A., Kandeler, E., Frossard, E. and Bünemann, E. K. (2019). Is the enzymatic hydrolysis of soil organic phosphorus compounds limited by enzyme or substrate availability? *Soil Biol. Biochem.* 139, 107628. doi:10.1016/j.soilbio.2019.107628.

Jarvie, H. P., Sharpley, A. N., Withers, P. J., Scott, J. T., Haggard, B. E. and Neal, C. (2013a). Phosphorus mitigation to control river eutrophication: murky waters, inconvenient truths, and "postnormal" science. *J. Environ. Qual.* 42(2), 295-304.

Jarvie, H. P., Sharpley, A. N., Spears, B., Buda, A. R., May, L. and Kleinman, P. J. A. (2013b). Water quality remediation faces unprecedented challenges from legacy phosphorus. *Environ. Sci. Technol.* 47(16), 8997-8998.

Jarvis, R. J. and Bolland, M. D. A. (1991). Lupin grain yields and fertiliser effectiveness are increased by banding superphosphate below the seed. *Aust. J. Exp. Agric.* 31(3), 357-366.

Jia, X., Liu, P. and Lynch, J. P. (2018). Greater lateral root branching density in maize (*Zea mays* L.) improves phosphorus acquisition from low phosphorus soil. *J. Exp. Bot.* 69(20), 4961–4970.

Johnston, A. E., Poulton, P. R. and White, R. P. (2013). Plant-available soil phosphorus. Part II: the response of arable crops to Olsen P on a sandy clay loam and a silty clay loam. *Soil Use Manag.* 29(1), 12–21.

Johnston, A. E., Poulton, P. R., Fixen, P. E. and Curtin, D. (2014). Phosphorus: its efficient use in agriculture. *Adv. Agron.* 123, 177–228.

Jones, D. L., Nguyen, C. and Finlay, R. D. (2009). Carbon flow in the rhizosphere: carbon trading at the soil-root interface. *Plant Soil* 321(1–2), 5–33.

Kandeler, E., Marschner, P., Tscherko, D., Gahoonia, T. S. and Nielsen, N. E. (2001). Microbial community composition and functional diversity in the rhizosphere of maize. *Plant Soil* 238, 301–312.

Kar, G., Peak, D. and Schoenau, J. J. (2012). Spatial distribution and chemical speciation of soil phosphorus in a band application. *Soil Sci. Soc. Am. J.* 76(6), 2297–2306.

Kidd, D. R., Ryan, M. H., Haling, R. E., Lambers, H., Sandral, G. A., Yang, Z., Culvenor, R. A., Cawthray, G. R., Stefanski, A. and Simpson, R. J. (2016). Rhizosphere carboxylates and morphological root traits in pasture legumes and grasses. *Plant Soil* 402(1–2), 77–89.

Klamer, F., Vogel, F., Li, X., Bremer, H., Neumann, G., Neuhäuser, B., Hochholdinger, F. and Ludewig, U. (2019). Estimating the importance of maize root hairs in low phosphorus conditions and under drought. *Ann. Bot.* 124(6), 961–968.

Kucey, R. M. N., Janzen, H. H. and Leggett, M. E. (1989). Microbially mediated increases in plant-available phosphorus. *Adv. Agron.* 42, 199–228.

Lambers, H., Shane, M. W., Cramer, M. D., Pearse, S. J. and Veneklaas, E. J. (2006). Root structure and functioning for efficient acquisition of phosphorus: matching morphological and physiological traits. *Ann. Bot.* 98(4), 693–713.

Lambers, H., Raven, J. A., Shaver, G. R. and Smith, S. E. (2008). Plant nutrient-acquisition strategies change with soil age. *Trends Ecol. Evol. (Amst.)* 23(2), 95–103.

Lambers, H., Brundrett, M. C., Raven, J. A. and Hopper, S. D. (2010). Plant mineral nutrition in ancient landscapes: high plant species diversity on infertile soils is linked to functional diversity for nutritional strategies. *Plant Soil* 334(1–2), 11–31.

Lean, G. R., Vizard, A. L. and Webb Ware, J. K. (1997). Changes in productivity and profitability of wool-growing farms that follow recommendations from agricultural and veterinary studies. *Aust. Vet. J.* 75(10), 726–731.

Leech, F. J., Richardson, A. E., Kertesz, M. A., Orchard, B. A., Banerjee, S. and Graham, P. (2019). Comparative effect of alternative fertilisers on pasture production, soil properties and soil microbial community structure. *Crop Pasture Sci.* 70(12), 1110–1127.

Lewis, C., Malcolm, B., Farquharson, R., Leury, B., Behrendt, R. and Clark, S. (2012). Economic analysis of improved perennial pasture systems. *Aust. Farm Bus. Manag. J.* 9, 37–55.

Lewis, D. C., Clarke, A. L. and Hall, W. B. (1981). Factors affecting the retention of phosphorus applied as superphosphate to the sandy soils in south-eastern South Australia. *Aust. J. Soil Res.* 19(2), 167–174.

Lewis, D. C., Clarke, A. L. and Hall, W. B. (1987). Accumulation of plant nutrients and changes in soil properties of sandy soils under fertilized pasture in southeastern South Australia. 1. Phosphorus. *Aust. J. Soil Res.* 25(2), 193–202.

Lewis, D. C., Bolland, M. D. A., Gilkes, R. J. and Hamilton, L. J. (1997). Review of Australian phosphate rock research. *Aust. J. Exp. Agric.* 37(8), 845–859.

Li, H., Ma, Q., Li, H., Zhang, F., Rengel, Z. and Shen, J. (2014a). Root morphological responses to localized nutrient supply differ among crop species with contrasting root traits. *Plant Soil* 376(1–2), 151–163.

Li, L., Zhang, F., Li, X., Christie, P., Sun, J., Yang, S. and Tang, C. (2003). Interspecific facilitation of nutrient uptake by intercropped maize and faba bean. *Nutr. Cycl. Agroecosyst.* 65(1), 61–71.

Li, L., Tilman, D., Lambers, H. and Zhang, F. S. (2014b). Plant diversity and overyielding: insights from belowground facilitation of intercropping in agriculture. *New Phytol.* 203(1), 63–69.

Loi, A., Howieson, J. G., Nutt, B. J. and Carr, S. J. (2005). A second generation of annual pasture legumes and their potential for inclusion in Mediterranean type farming systems. *Aust. J. Exp. Agric.* 45(3), 289–299.

Lombi, E., McLaughlin, M. J., Johnston, C., Armstrong, R. D. and Holloway, R. E. (2004). Mobility and lability of phosphorus from granular and fluid monoammonium phosphate differs in a calcareous soil. *Soil Sci. Soc. Am. J.* 68(2), 682–689.

Lun, F., Liu, J., Ciais, P., Nesme, T., Chang, J., Wang, R., Goll, D., Sardans, J., Peñuelas, J. and Obersteiner, M. (2018). Global and regional phosphorus budgets in agricultural systems and their implications for phosphorus-use efficiency. *Earth Syst. Sci. Data* 10(1), 1–18.

Lynch, J. P. (2007). Roots of the second green revolution. *Aust. J. Bot.* 55(5), 493–512.

Lynch, J. P. (2011). Root phenes for enhanced soil exploration and phosphorus acquisition: tools for future crops. *Plant Physiol.* 156(3), 1041–1049.

Lynch, J. P. (2019). Root phenotypes for improved nutrient capture: an underexploited opportunity for global agriculture. *New Phytol.* 223(2), 548–564.

Lyu, Y., Tang, H., Li, H., Zhang, F., Rengel, Z., Whalley, W. R. and Shen, J. (2016). Major crop species show differential balance between root morphological and physiological responses to variable phosphorus supply. *Front. Plant Sci.* 7, 1939. doi: 10.3389/fpls.2016.01939.

MacDonald, G. K., Bennett, E. M., Potter, P. A. and Ramankutty, N. (2011). Agronomic phosphorus imbalances across the world's croplands. *Proc. Natl Acad. Sci. U.S.A.* 108(7), 3086–3091.

Mai, W., Xue, X., Feng, G., Yang, R. and Tian, C. (2018). Can optimization of phosphorus input lead to high productivity and high phosphorus use efficiency of cotton through maximization of root/mycorrhizal efficiency in phosphorus acquisition? *Field Crops Res.* 216, 100–108.

Mai, W., Xue, X., Feng, G., Yang, R. and Tian, C. (2019). Arbuscular mycorrhizal fungi - 15-fold enlargement of the soil volume of cotton roots for phosphorus uptake in intensive planting conditions. *Eur. J. Soil Biol.* 90, 31–35.

Malcolm, B., Smith, K. F. and Jacobs, J. L. (2014). Perennial pasture persistence: the economic perspective. *Crop Pasture Sci.* 65(8), 713–720.

Manske, G. G. B., Ortiz-Monasterio, J. I., Van Ginkel, M., González, R. M., Rajaram, S., Molina, E. and Vlek, P. L. G. (2000). Traits associated with improved P-uptake efficiency in CIMMYT's semidwarf spring bread wheat grown on an acid andisol in Mexico. *Plant Soil* 221(2), 189–204.

Marcus, O. O. (2013). Genetic assessment and mapping of QTLs for biomass, P uptake, crown root angle and mycorrhiza colonization for adaptation to low phosphorus conditions in west African *Sorghum bicolor* L. Master Thesis, University of Hohenheim, Germany.

Marschner, P., Marhan, S. and Kandeler, E. (2012). Microscale distribution and function of soil microorganisms in the interface between rhizosphere and detritusphere. *Soil Biol. Biochem.* 49, 174–183.

Mat Hassan, H., Hasbullah, H. and Marschner, P. (2013). Growth and rhizosphere P pools of legume–wheat rotations at low P supply. *Biol. Fertil. Soils* 49(1), 41–49.

McBeath, T. M., McLaughlin, M. J., Kirby, J. K. and Armstrong, R. D. (2012). The effect of soil water status on fertiliser, topsoil and subsoil phosphorus utilisation by wheat. *Plant Soil* 358(1–2), 337–348.

McCaskill, M. R. and Cayley, J. W. D. (2000). Soil audit of a long-term phosphate experiment in south-western Victoria: total phosphorus, sulfur, nitrogen, and major cations. *Aust. J. Agric. Res.* 51(6), 737–748.

McDowell, R. W., Sharpley, A. N., Condron, L. M., Haygarth, P. M. and Brookes, P. C. (2001). Processes controlling soil phosphorus release to runoff and implications for agricultural management. *Nutr. Cycl. Agroecosyst.* 59(3), 269–284.

McEachern, S. F. J. and Brown, D. (2010). *AgInsights 2009*. Wagga Wagga, Australia: Holmes Sackett Pty Ltd.

McLachlan, J. W., Haling, R. E., Simpson, R. J., Li, X., Flavel, R. J. and Guppy, C. N. (2019). Variation in root morphology and P-acquisition efficiency among *Trifolium subterraneum* genotypes. *Crop Pasture Sci.* 70(11), 1015–1032.

McLachlan, J. W., Becquer, A., Haling, R. E., Simpson, R. J., Flavel, R. J. and Guppy, C. N. (2020). Intrinsic root morphology determines the phosphorus acquisition efficiency of five annual pasture legumes irrespective of mycorrhizal colonisation. *Funct. Plant Biol.* https://doi.org/10.1071/FP20007.

McLachlan, K. D. (1965). The nature of available phosphorus in some acid pasture soils and a comparison of estimating procedures. *Aust. J. Exp. Agric.* 5(17), 125–132.

McLaren, T. I., Simpson, R. J., McLaughlin, M. J., Smernik, R. J., McBeath, T. M., Guppy, C. N. and Richardson, A. E. (2015). An assessment of various measures of soil phosphorus and the net accumulation of phosphorus in fertilized soils under pasture. *J. Plant Nutr. Soil Sci.* 178(4), 543–554.

McLaren, T. I., McBeath, T. M., Simpson, R. J., Richardson, A. E., Stefanski, A., Guppy, C. N., Smernik, R. J., Rivers, C., Johnston, C. and McLaughlin, M. J. (2017a). Direct recovery of ^{33}P-labelled fertiliser phosphorus in subterranean clover (*Trifolium subterraneum*) pastures under field conditions – the role of agronomic management. *Agric. Ecosyst. Environ.* 246, 144–156.

McLaren, T. I., Smernik, R. J., Simpson, R. J., McLaughlin, M. J., McBeath, T. M., Guppy, C. N. and Richardson, A. E. (2017b). The chemical nature of organic phosphorus that accumulates in fertilized soils of a temperate pasture as determined by solution ^{31}P NMR spectroscopy. *J. Plant Nutr. Soil Sci.* 180(1), 27–38.

McLaren, T. I., Smernik, R. J., McLaughlin, M. J., McBeath, T. M., McCaskill, M. R., Robertson, F. A. and Simpson, R. J. (2020). Soil phosphorus pools with addition of fertiliser phosphorus in a long-term grazing experiment. *Nutr. Cycl. Agroecosyst.* 116(2), 151–164.

McLaughlin, M. J., Alston, A. M. and Martin, J. K. (1988). Phosphorus cycling in wheat pasture rotations. I. The source of phosphorus taken up by wheat. *Aust. J. Soil Res.* 26(2), 323–331.

McLaughlin, M. J., Baker, T. G., James, T. R. and Rundle, J. A. (1990). Distribution and forms of phosphorus and aluminum in acidic topsoils under pastures in south-eastern Australia. *Aust. J. Soil Res.* 28(3), 371–385.

McLaughlin, M. J., McBeath, T. M., Smernik, R., Stacey, S. P., Ajiboye, B. and Guppy, C. (2011). The chemical nature of P-accumulation in agricultural soils - implications for fertiliser management and design: an Australian perspective. *Plant Soil* 349(1-2), 69–87.

Melland, A. R., McCaskill, M. R., White, R. E. and Chapman, D. F. (2008). Loss of phosphorus and nitrogen in runoff and subsurface drainage from high and low input pastures grazed by sheep in Australia. *Aust. J. Soil Res.* 46, 161–172.

Meyer, G., Maurhofer, M., Frossard, E., Gamper, H. A., Mäder, P., Mészáros, É., Schönholzer-Mauclaire, L., Symanczik, S. and Oberson, A. (2019). *Pseudomonas protegens* CHA0 does not increase phosphorus uptake from ^{33}P labeled synthetic hydroxyapatite by wheat grown on calcareous soil. *Soil Biol. Biochem.* 131, 217–228.

Miguel, M. A., Widrig, A., Vieira, R. F., Brown, K. M. and Lynch, J. P. (2013). Basal root whorl number: a modulator of phosphorus acquisition in common bean (*Phaseolus vulgaris*). *Ann. Bot.* 112(6), 973–982.

Miguel, M. A., Postma, J. A. and Lynch, J. P. (2015). Phene synergism between root hair length and basal root growth angle for phosphorus acquisition. *Plant Physiol.* 167(4), 1430–1439.

Mills, A., Moot, D. J. and McKenzie, B. A. (2006). Cocksfoot pasture production in relation to environmental variables. *Proc. N. Z. Grassland Assoc.* 68, 89–94.

Mitsukawa, N., Okumura, S., Shirano, Y., Sato, S., Kato, T., Harashima, S. and Shibata, D. (1997). Overexpression of an *Arabidopsis thaliana* high-affinity phosphate transporter gene in tobacco cultured cells enhances cell growth under phosphate-limited conditions. *Proc. Natl Acad. Sci. U.S.A.* 94(13), 7098–7102.

Moody, P. W. (2007). Interpretation of a single-point P buffering index for adjusting critical levels of the Colwell soil P test. *Aust. J. Soil Res.* 45(1), 55–62.

Moore, R. M. (1970a). *Australian Grasslands*. Canberra, Australia: Australian National University Press.

Moore, R. M. (1970b). Australian grasslands. In: Moore, R. M. (Ed.), *Australian Grasslands*. Canberra, Australia: Australian National University Press, pp. 85–100.

Mori, A., Fukuda, T., Vejchasarn, P., Nestler, J., Pariasca-Tanaka, J. and Wissuwa, M. (2016). The role of root size versus root efficiency in phosphorus acquisition in rice. *J. Exp. Bot.* 67(4), 1179–1189.

Morley, F. H. W. (1961). Subterranean clover. *Adv. Agron.* 13, 57–123.

Nestler, J. and Wissuwa, M. (2016). Superior root hair formation confers root efficiency in some, but not all, rice genotypes upon P deficiency. *Front. Plant Sci.* 7, 1935. doi:10.3389/fpls.2016.01935.

Nichols, P. G. H., Loi, A., Nutt, B. J., Evans, P. M., Craig, A. D., Pengelly, B. C., Dear, B. S., Lloyd, D. L., Revell, C. K., Nair, R. M., Ewing, M. A., Howieson, J. G., Auricht, G. A., Howie, J. H., Sandral, G. A., Carr, S. J., de Koning, C. T., Hackney, B. F., Crocker, G. J., Snowball, R., Hughes, S. J., Hall, E. J., Foster, K. J., Skinner, P. W., Barbetti, M. J. and You, M. P. (2007). New annual and short-lived perennial pasture legumes for Australian agriculture - 15 years of revolution. *Field Crops Res.* 104(1-3), 10–23.

Nichols, P. G. H., Revell, C. K., Humphries, A. W., Howie, J. H., Hall, E. J., Sandral, G. A., Ghamkhar, K. and Harris, C. A. (2012). Temperate pasture legumes in Australia - their history, current use, and future prospects. *Crop Pasture Sci.* 63(9), 691–725.

Nichols, P. G. H., Foster, K. J., Piano, E., Pecetti, L., Kaur, P., Ghamkhar, K. and Collins, W. J. (2013). Genetic improvement of subterranean clover (*Trifolium subterraneum* L.). 1. Germplasm, traits and future prospects. *Crop Pasture Sci.* 64(4), 312–346.

Nicol, D. L., Finlayson, J., Colmer, T. D. and Ryan, M. H. (2013). Opportunistic Mediterranean agriculture - Using ephemeral pasture legumes to utilize summer rainfall. *Agric. Syst.* 120, 76–84.

Nutt, B. J. (2004a). French serradella (*Ornithopus sativus*) 'Erica'. *Plant Var. J.* 17, 313–315.

Nutt, B. J. (2004b). French serradella (*Ornithopus sativus*) 'Margurita'. *Plant Var. J.* 17, 310–312.

Oburger, E., Jones, D. L. and Wenzel, W. W. (2011). Phosphorus saturation and pH differentially regulate the efficiency of organic acid anion-mediated P solubilization mechanisms is soil. *Plant Soil* 341(1–2), 363–382.

Oehl, F., Oberson, A., Probst, M., Fliessbach, A., Roth, H. and Frossard, E. (2001). Kinetics of microbial phosphorus uptake in cultivated soils. *Biol. Fertil. Soils* 34(1), 31–41.

Olsen, S. R., Cole, C. V., Watanabe, F. S. and Dean, L. A. (1954). *Estimation of Available Phosphorus in Soils by Extraction with Sodium Bicarbonate*. Washington, DC: United States Department of Agriculture, Circular No. 939.

Osman, A. E., Cocks, P. S., Russi, L. and Pagnotta, M. A. (1991). Response of Mediterranean grassland to phosphate and stocking rates: biomass production and botanical composition. *J. Agric. Sci.* 116(1), 37–46.

Ozanne, P. G. (1980). Phosphate nutrition of plants – A general treatise. In: Khasawneh, F. E., Sample, E. C. and Kamprath, E. J. (Eds), *The Role of Phosphorus in Agriculture*. Madison, WI: American Society of Agronomy/Crop Science Society of America/Soil Science Society of America, pp. 559–589.

Ozanne, P. G., Kirton, D. J. and Shaw, T. C. (1961). The loss of phosphorus from sandy soils. *Aust. J. Agric. Res.* 12(3), 409–423.

Ozanne, P. G. and Shaw, T. C. (1967). Phosphate sorption by soils as a measure of the phosphate requirement for pasture growth. *Aust. J. Agric. Res.* 18(4), 601–612.

Ozanne, P. G., Keay, J. and Biddiscombe, E. F. (1969). The Comparative applied phosphate requirements of eight annual pasture species. *Aust. J. Agric. Res.* 20(5), 809–818.

Pang, J., Tibbett, M., Denton, M. D., Lambers, H., Siddique, K. H. M., Bolland, M. D. A., Revell, C. K. and Ryan, M. H. (2010). Variation in seedling growth of 11 perennial legumes in response to phosphorus supply. *Plant Soil* 328(1–2), 133–143.

Paynter, B. H. (1990). Comparative phosphate requirements of yellow serradella (*Ornithopus compressus*), burr medic (*Medicago polymorpha* var *brevispina*) and subterranean clover (*Trifolium subterraneum*). *Aust. J. Exp. Agric.* 30(4), 507–514.

Peoples, M. B., Brockwell, J., Hunt, J. R., Swan, A. D., Watson, L., Hayes, R. C., Li, G. D., Hackney, B., Nuttall, J. G., Davies, S. L. and Fillery, I. R. P. (2012). Factors affecting the potential contributions of N_2 fixation by legumes in Australian pasture systems. *Crop Pasture Sci.* 63(9), 759–786.

Perrott, K. W., Sarathchandra, S. U. and Waller, J. E. (1990). Seasonal storage and release of phosphorus and potassium by organic matter and the microbial biomass in a high-producing pastoral soil. *Aust. J. Soil Res.* 28(4), 593–608.

Perrott, K. W., Sarathchandra, S. U. and Dow, B. W. (1992). Seasonal and fertilizer effects on the organic cycle and microbial biomass in a hill country soil under pasture. *Soil Biol. Biochem.* 30(3), 383–394.

Pinkerton, A. and Simpson, J. R. (1986). Interactions of surface drying and subsurface nutrients affecting plant growth on acidic soil profiles from an old pasture. *Aust. J. Exp. Agric.* 26(6), 681–689.

Postma, J. A. and Lynch, J. P. (2011). Root cortical aerenchyma enhances the growth of maize on soils with suboptimal availability of nitrogen, phosphorus, and potassium. *Plant Physiol.* 156(3), 1190–1201.

Poulton, P. R., Johnston, A. E. and White, R. P. (2013). Plant-available soil phosphorus. Part I: the response of winter wheat and spring barley to Olsen P on a silty clay loam. *Soil Use Manag.* 29(1), 4–11.

Probert, M. E. and Jones, R. K. (1977). The use of soil analysis for predicting the response to sulphur of pasture legumes in the Australian tropics. *Aust. J. Soil Res.* 15(2), 137–146.

Rae, A. L., Jarmey, J. M., Mudge, S. R. and Smith, F. W. (2004). Over-expression of a high-affinity phosphate transporter in transgenic barley plants does not enhance phosphate uptake rates. *Funct. Plant Biol.* 31(2), 141–148.

Reuter, D. J., Dyson, C. B., Elliott, D. E., Lewis, D. C. and Rudd, C. L. (1995). An appraisal of soil phosphorus testing data for crops and pastures in South Australia. *Aust. J. Exp. Agric.* 35(7), 979–995.

Revell, C. and Revell, D. (2007). Meeting 'duty of care' obligations when developing new pasture species. *Field Crops Res.* 104(1–3), 95–102.

Richardson, A. E. (2009). Regulating the phosphorus nutrition of plants: molecular biology meeting agronomic needs. *Plant Soil* 322(1–2), 17–24.

Richardson, A. E., Hadobas, P. A., Hayes, J. E., O'Hara, C. P. and Simpson, R. J. (2001). Utilization of phosphorus by pasture plants supplied with myo-inositol hexaphosphate is enhanced by the presence of soil microorganisms. *Plant Soil* 229(1), 47–56.

Richardson, A. E., Lynch, J. P., Ryan, P. R., Delhaize, E., Smith, F. A., Smith, S. E., Harvey, P. R., Ryan, M. H., Veneklaas, E. J., Lambers, H., Oberson, A., Culvenor, R. A. and Simpson, R. J. (2011). Plant and microbial strategies to improve the phosphorus efficiency of agriculture. *Plant Soil* 349(1–2), 121–156.

Richardson, A. E. and Simpson, R. J. (2011). Soil microorganisms mediating phosphorus availability update on microbial phosphorus. *Plant Physiol.* 156(3), 989–996.

Ridley, A. M., White, R. E., Simpson, R. J. and Callinan, L. (1997). Water use and drainage under phalaris, cocksfoot, and annual ryegrass pastures. *Aust. J. Agric. Res.* 48(7), 1011–1023.

Ridley, A. M., Christy, B. P., White, R. E., McLean, T. and Green, R. (2003). North-East Victoria SGS National Experiment site: water and nutrient losses from grazing systems on contrasting soil types and levels of inputs. *Aust. J. Exp. Agric.* 43(8), 799–815.

Ritchie, G. S. P. and Weaver, D. M. (1993). Phosphorus retention and release from sandy soils of the Peel-Harvey catchment. *Fert. Res.* 36(2), 115–122.

Roberts, T. L. and Johnston, A. E. (2015). Phosphorus use efficiency and management in agriculture. *Resour. Conserv. Recycl.* 105, 275–281.

Robinson, D. and Van Vuuren, M. M. I. (1998). Responses of wild plants to nutrient patches in relation to growth rate and life-form. In: Lambers, H., Poorter, H. and Van Vuuren, M. M. I. (Eds), *Inherent Variation in Plant Growth. Physiological Mechanisms and Ecological Consequences.* Leiden: Backhuys Publishers, pp. 237–257.

Robinson, K., Bell, L. W., Bennett, R. G., Henry, D. A., Tibbett, M. and Ryan, M. H. (2007). Perennial legumes native to Australia – a preliminary investigation of nutritive value and response to cutting. *Aust. J. Exp. Agric.* 47(2), 170–176.

Rose, T. J. and Wissuwa, M. (2012). Rethinking internal phosphorus utilization efficiency: a new approach is needed to improve PUE in grain crops. *Adv. Agron.* 116, 185-217.

Rose, T. J., Liu, L. and Wissuwa, M. (2013). Improving phosphorus efficiency in cereal crops: is breeding for reduced grain phosphorus concentration part of the solution? *Front. Plant. Sci.* 4, 444.

Rossiter, R. C. (1966). Ecology of the Mediterranean annual type pasture. *Adv. Agron.* 18, 1-56.

Rowe, H., Withers, P. J. A., Baas, P., Chan, N. I., Doody, D., Holiman, J., Jacobs, B., Li, H., MacDonald, G. K., McDowell, R., Sharpley, A. N., Shen, J., Taheri, W., Wallenstein, M. and Weintraub, M. N. (2016). Integrating legacy soil phosphorus into sustainable nutrient management strategies for future food, bioenergy and water security. *Nutr. Cycl. Agroecosyst.* 104(3), 393-412.

Rudd, C. L. (1972). Response of annual medic pasture to superphosphate applications and correlation with available soil phosphorus. *Aust. J. Exp. Agric. Anim. Husb.* 12, 43-48.

Russell, J. S. (1960). Soil fertility changes in the long term experimental plots at Kybybolite, South Australia. II. Changes I phosphorus. *Aust. J. Agric. Res.* 11(6), 926-947.

Ryan, M. H. and Graham, J. H. (2018). Little evidence that farmers should consider abundance or diversity of arbuscular mycorrhizal fungi when managing crops. *New Phytol.* 220(4), 1092-1107.

Ryan, P., Delhaize, E. and Jones, D. (2001). Function and mechanisms of organic anion exudation from roots. *Annu. Rev. Plant. Physiol. Plant. Mol. Biol.* 52(1), 527-560.

Ryan, P. R., James, R. A., Weligama, C., Delhaize, E., Rattey, A., Lewis, D. C., Bovill, W. D., McDonald, G., Rathjen, T. M., Wang, E., Fettell, N. A. and Richardson, A. E. (2014). Can citrate efflux from roots improve phosphorus uptake by plants? Testing the hypothesis with near-isogenic lines of wheat. *Physiol. Plant.* 151(3), 230-242.

Sale, P. W. G., Gilkes, R. J., Bolland, M. D. A., Simpson, P. G., Lewis, D. C., Ratkowsky, D. A., Gilbert, M. A., Garden, D. L., Cayley, J. W. D. and Johnson, D. (1997). The agronomic effectiveness of reactive phosphate rocks. 1. Effect of the pasture environment. *Aust. J. Exp. Agric.* 37(8), 921-936.

Sample, E. C., Soper, R. J. and Racz, G. J. (1980). Reactions of phosphate fertilizers in soils. In: Khasawneh, F. E., Sample, E. C. and Kamprath, E. J. (Eds), *The Role of Phosphorus in Agriculture*. Madison WI: American Society of Agronomy, Crop Science Society of America, Soil Science Society of America, pp. 263-310.

Sanchez, C. A. , Porter, P. S. and Ulloa, M. F. (1991). Relative efficiency of broadcast and banded phosphorus for sweet corn produced on histosols. *Soil Sci. Soc. Am. J.* 55(3), 871-875.

Sandral, G. A., Haling, R. E., Ryan, M. H., Price, A., Pitt, W. M., Hildebrand, S. M., Fuller, C. G., Kidd, D. R., Stefanksi, A., Lambers, H. and Simpson, R. J. (2018). Intrinsic capacity for nutrient foraging predicts critical external phosphorus requirement of 12 pasture legumes. *Crop Pasture Sci.* 69(2), 174-182.

Sandral, G. A., Price, A., Hildebrand, S. M., Fuller, C. G., Haling, R. E., Stefanski, A., Yang, Z., Culvenor, R. A., Ryan, M. H., Kidd, D. R., Diffey, S., Lambers, H. and Simpson, R. J. (2019). Field benchmarking of the critical external phosphorus requirements of pasture legumes for southern Australia. *Crop Pasture Sci.* 70(12), 1080-1096.

Sattari, S. Z., Bouwman, A. F., Giller, K. E. and Van Ittersum, M. K. (2012). Residual soil phosphorus as the missing link in the global phosphorus crisis puzzle. *Proc. Natl Acad. Sci. U.S.A.* 109(16), 6348-6353.

Sattari, S. Z., van Ittersum, M. K., Bouwman, A. F., Smit, A. L. and Janssen, B. H. (2014). Crop yield response to soil fertility and N, P, K inputs in different environments: testing and improving the QUEFTS model. *Field Crops Res.* 157, 35–46.

Schefe, C. R., Barlow, K. M., Robinson, N. J., Crawford, D. M., McLaren, T. I., Smernik, R. J., Croatto, G., Walsh, R. D. and Kitching, M. (2015). 100 years of superphosphate addition to pasture in an acid soil – current nutrient status and future management. *Soil Res.* 53(6), 662–676.

Schneider, H. M., Postma, J. A., Wojciechowski, T., Kuppe, C. and Lynch, J. P. (2017a). Root cortical senescence improves growth under suboptimal availability of N, P, and K. *Plant Physiol.* 174(4), 2333–2347.

Schneider, K. D., Voroney, R. P., Lynch, D. H., Oberson, A., Frossard, E. and Bünemann, E. K. (2017b). Microbially-mediated P fluxes in calcareous soils as a function of water-extractable phosphate. *Soil Biol. Biochem.* 106, 51–60.

Scholz, R. W. and Wellmer, F.-W. (2015). Losses and use efficiencies along the phosphorus cycle. Part 1: Delemmata and losses on the mines and other nodes of the supply chain. *Resour. Conserv. Recycl.* 105, 216–234.

Scholz, R. W. and Wellmer, F.-W. (2018). Although there is no physical short-term scarcity of phosphorus, its resource efficiency should be improved. *J. Indust. Ecol.* 23(2), 313–318.

Schweiger, P. F., Robson, A. D. and Barrow, N. J. (1995). Root hair length determines beneficial effect of a *Glomus* species on shoot growth of some pasture species. *New Phytol.* 131(2), 247–254.

Scott, B. J. (1973). The response of barrel medic pasture to topdressed and placed superphosphate in central western New South Wales. *Aust. J. Exp. Agric.* 13(65), 705–710.

Scott, B. J., Conyers, M. K., Poile, G. J. and Cullis, B. R. (1997). Subsurface acidity and liming affect yield of cereals. *Aust. J. Agric. Res.* 48(6), 843–854.

Scott, B. J., Fisher, J. A. and Cullis, B. R. (2001). Aluminium tolerance and lime increase wheat yield on the acidic soils of central and southern New South Wales. *Aust. J. Exp. Agric.* 41(4), 523–532.

Scott, B. J. and Coombes, N. E. (2006). Poor incorporation of lime limits grain yield response in wheat. *Aust. J. Exp. Agric.* 46(11), 1481–1487.

Seeling, B. and Zasoski, R. J. (1993). Microbial effects in maintaining organic and inorganic solution phosphorus concentrations in a grassland topsoil. *Plant Soil* 148(2), 277–284.

Sharpley, A. N. (1995). Dependence of runoff phosphorus on extractable soil phosphorus. *J. Environ. Qual.* 24(5), 920–926.

Simpson, J. R., Bromfield, S. M. and Jones, L. (1974). Effects of management on soil fertility under pasture. 3. Changes in total soil nitrogen, carbon, phosphorus and exchangeable cations. *Aust. J. Exp. Agric.* 14(69), 487–494.

Simpson, R., Graham, P., Davies, L. and Zurcher, E. (2009). Five easy steps to ensure you are making money from superphosphate. Canberra and Orange, Australia: CSIRO & Industry and Investment NSW. Available at: https://www.mla.com.au/globalassets/mla-corporate/generic/extension-training-and-tools/5-easy-steps-guide.pdf.

Simpson, R., Haling, R., Virgona, J. and Ferguson, N. (2017). Managing the phosphorus cycle in clover-based pasture for more effective use of P-fertiliser inputs. *Proceedings of the 58th Ann. Conf. Grassld Soc. Southern Aust.* Carrum Downs, Victoria, Australia: Grassland Society of Southern Australia Inc., pp. 17–21.

Simpson, R. J., Oberson, A., Culvenor, R. A., Ryan, M. H., Veneklaas, E. J., Lambers, H., Lynch, J. P., Ryan, P. R., Delhaize, E., Smith, F. A., Smith, S. E., Harvey, P. R. and Richardson, A. E. (2011a). Strategies and agronomic interventions to improve the phosphorus-use efficiency of farming systems. *Plant Soil* 349(1–2), 89–120.

Simpson, R. J., Richardson, A. E., Riley, I. T., McKay, A. C., McKay, S. F., Ballard, R. A., Ophel-Keller, K., Hartley, D., O'Rourke, T. A., Li, H., Sivasithamparam, K., Ryan, M. H. and Barbetti, M. J. (2011b). Damage to roots of *Trifolium subterraneum* L. (subterranean clover), failure of seedlings to establish and the presence of root pathogens during autumn–winter. *Grass Forage Sci.* 66(4), 585–605.

Simpson, R. J., Richardson, A. E., Nichols, S. N. and Crush, J. R. (2014). Pasture plants and soil fertility management to improve the efficiency of phosphorus fertiliser use in temperate grassland systems. *Crop Pasture Sci.* 65(6), 556–575.

Simpson, R. J., Stefanski, A., Marshall, D. J., Moore, A. D. and Richardson, A. E. (2015). Management of soil phosphorus fertility determines the phosphorus budget of a temperate grazing system and is the key to improving phosphorus-balance efficiency. *Agric. Ecosyst. Environ.* 212, 263–277.

Slattery, W. J., Conyers, M. K. and Aitken, R. L. (1999). Soil pH, aluminium, manganese and lime requirement. In: Peverill, K. I., Sparrow, L. A. and Reuter, D. J. (Eds), *Soil Analysis: An Interpretation Manual*. Collingwood, Australia: CSIRO Publishing, pp. 103–128.

Smith, D. F. (2000). *Natural Gain: In the Grazing Lands of Southern Australia*. Sydney, Australia: University of New South Wales Press.

Smith, L. C., Moss, R. A., Morton, J. D., Metherell, A. K. and Fraser, T. J. (2012). Pasture production from a long-term fertiliser trial under irrigation. *N. Z. J. Agric. Res.* 55(2), 105–117.

Smith, S. E. and Read, D. J. (2008). *Mycorrhizal Symbiosis*. London: Academic Press.

Smith, S. E., Anderson, I. C. and Smith, F. A. (2015). Mycorrhizal associations and phosphorus acquisition: from cells to ecosystems. *Ann. Plant Rev.* 48, 409–440.

Stewart, W. M., Dibb, D. W., Johnston, A. E. and Smyth, T. J. (2005). The contribution of commercial fertilizer nutrients to food production. *Agron. J.* 97(1), 1–6.

Strock, C. F., de la Riva, L. M. and Lynch, J. P. (2018). Reduction in root secondary growth as a strategy for phosphorus acquisition. *Plant Physiol.* 176(1), 691–703.

Sun, B., Gao, Y. and Lynch, J. P. (2018). Large crown root number improves topsoil foraging and phosphorus acquisition. *Plant Physiol.* 177(1), 90–104.

Sutton, M. A., Bleeker, A., Howard, C. M., Bekunda, M., Grizzetti, B., de Vries, W., van Grinsven, H. J. M., Abrol, Y. P., Adhya, T. K., Billen, G., Davidson, E. A., Datta, A., Diaz, R., Erisman, J. W., Liu, X. J., Oenema, O., Palm, C., Raghuram, N., Reis, S., Scholz, R. W., Sims, T., Westhoek, H. and Zhang, F. S., with contributions from Ayyappan, S., Bouwman, A. F., Bustamante, M., Fowler, D., Galloway, J. N., Gavito, M. E., Garnier, J., Greenwood, S., Hellums, D. T., Holland, M., Hoysall, C., Jaramillo, V. J., Klimont, Z., Ometto, J. P., Pathak, H., Plocq Fichelet, V., Powlson, D., Ramakrishna, K., Roy, A., Sanders, K., Sharma, C., Singh, B., Singh, U., Yan, X. Y. and Zhang, Y. (2013). Our Nutrient World: the challenge to produce more food and energy with less pollution. In: *Global Overview of Nutrient Management*. Centre for Ecology and Hydrology, Edinburgh on behalf of the Global Partnership on Nutrient Management and the International Nitrogen Initiative. Available at: https://www.unenvironment.org/resou rces/report/our-nutrient-world-challenge-produce-more-food-and-energy-less-pollution.

Svenningsen, N. B., Watts-Williams, S. J., Joner, E. J., Battini, F., Efthymiou, A., Cruz-Paredes, C., Nybroe, O. and Jakobsen, I. (2018). Suppression of the activity of arbuscular mycorrhizal fungi by the soil microbiota. *ISME J.* 12(5), 1296-1307.

Tarafdar, J. C. and Jungk, A. (1987). Phosphatase activity in the rhizosphere and its relation to the depletion of soil organic phosphorus. *Biol. Fertil. Soils* 3(4), 199-204.

Tinker, P. B. and Nye, P. H. (2000). *Solute Movement in the Rhizosphere*. New York: Oxford University Press Inc.

Tilman, D., Balzer, C., Hill, J. and Befort, B. L. (2011). Global food demand and the sustainable intensification of agriculture. *Proc. Natl Acad. Sci. U.S.A.* 108(50), 20260-20264.

Tonini, D., Saveyn, H. G. M. and Huygens, D. (2019). Environmental and health co-benefits for advanced phosphorus recovery. *Nat. Sustainability* 2(11), 1051-1061.

Trotter, M., Guppy, C., Haling, R., Trotter, T., Edwards, C. and Lamb, D. (2014). Spatial variability in pH and key soil nutrients: is this an opportunity to increase fertiliser and lime-use efficiency in grazing systems? *Crop Pasture Sci.* 65(8), 817-827.

Vaccari, D. A., Powers, S. M. and Liu, X. (2019). Demand-driven model for global phosphate rock suggests paths for phosphorus sustainability. *Environ. Sci. Technol.* 53(17), 10417-10425.

Valizadeh, G. R., Rengel, Z. and Rate, A. W. (2003). Response of wheat genotypes efficient in P utilization and genotypes responsive to P fertilization to different P banding depths and watering regimes. *Aust. J. Agric. Res.* 54(1), 59-65.

Vance, C. P., Uhde-Stone, C. and Allan, D. L. (2003). Phosphorus acquisition and use: critical adaptations by plants for securing a non-renewable resource. *New Phytol.* 157(3), 423-447.

Vandamme, E., Renkens, M., Pypers, P., Smolders, E., Vanlauwe, B. and Merckx, R. (2013). Root hairs explain P uptake efficiency of soybean genotypes grown in a P-deficient ferralsol. *Plant Soil* 369(1-2), 269-282.

Veneklaas, E. J., Lambers, H., Bragg, J., Finnegan, P. M., Lovelock, C. E., Plaxton, W. C., Price, C. A., Scheible, W. R., Shane, M. W., White, P. J. and Raven, J. A. (2012). Opportunities for improving phosphorus-use efficiency in crop plants. *New Phytol.* 195(2), 306-320.

Vitousek, P. M., Naylor, R., Crews, T., David, M. B., Drinkwater, L. E., Holland, E., Johnes, P. J., Katzenberger, J., Martinelli, L. A., Matson, P. A., Nziguheba, G., Ojima, D., Palm, C. A., Robertson, G. P. , Sanchez, P. A., Townsend, A. R. and Zhang, F. S. (2009). Nutrient imbalances in agricultural development. *Science* 324(5934), 1519-1520.

Waddell, H. A., Simpson, R. J., Henderson, B., Ryan, M. H., Lambers, H., Garden, D. L. and Richardson, A. E. (2015). Differential growth response of *Rytidosperma* species (wallaby grass) to phosphorus application and implications for grassland management. *Grass Forage Sci.* 71, 245-258.

Wang, Y., Krogstad, T., Clarke, J. L., Hallama, M., Øgaard, A. F., Eich-Greatorex, S., Kandeler, E. and Clarke, N. (2016). Rhizosphere organic anions play a minor role in improving crop species' ability to take up residual phosphorus (P) in agricultural soils low in P availability. *Front. Plant Sci.* 7, 1664. doi:10.3389/fpls.2016.01664.

Wang, Y. and Lambers, H. (2020). Root-released organic anions in response to low phosphorus availability: recent progress, challenges and future perspectives. *Plant Soil* 447(1-2), 135-156.

Watt, M., Silk, W. K. and Passioura, J. B. (2006). Rates of root and organism growth, soil conditions, and temporal and spatial development of the rhizosphere. *Ann. Bot.* 97(5), 839–855.

Weaver, D. M. and Wong, M. T. F. (2011). Scope to improve phosphorus (P) management and balance efficiency of crop and pasture soils with contrasting P status and buffering indices. *Plant Soil* 349(1–2), 37–54.

Weber, N. F., Herrmann, I., Hochholdinger, F., Ludewig, U. and Neumann, G. (2018). PGPR-induced growth stimulation and nutrient acquisition in maize: do root hairs matter? *Sci. Agric. Bohem.* 49(3), 164–172.

Wen, T. J. and Schnable, P. S. (1994). Analyses of mutants of three genes that influence root hair development in *Zea mays* (Gramineae) suggest that root hairs are dispensable. *Am. J. Bot.* 81(7), 833–842.

Wen, Z., Li, H., Shen, Q., Tang, X., Xiong, C., Li, H., Pang, J., Ryan, M. H., Lambers, H. and Shen, J. (2019). Tradeoffs among root morphology, exudation and mycorrhizal symbioses for phosphorus-acquisition strategies of 16 crop species. *New Phytol.* 223(2), 882–895.

Williams, C. H. and Andrew, C. S. (1970). Mineral nutrition of pastures. In: Moore, R. M. (Ed.), *Australian Grasslands*. Canberra, Australia: Australian National University Press, pp. 321–338.

Wissuwa, M., Kretzschmar, T. and Rose, T. J. (2016). From promise to application: root traits for enhanced nutrient capture in rice breeding. *J. Exp. Bot.* 67(12), 3605–3615.

Withers, P. J. A., Vadas, P. A., Uusitalo, R., Forber, K. J., Hart, M., Foy, R. H., Delgado, A., Dougherty, W., Lilja, H., Burkitt, L. L., Rubæk, G. H., Pote, D., Barlow, K., Rothwell, S. and Owens, P. R. (2019). A global perspective on integrated strategies to manage soil phosphorus status for eutrophication control without limiting land productivity. *J. Environ. Qual.* 48(5), 1234–1246.

Yang, Z., Culvenor, R. A., Haling, R. E., Stefanski, A., Ryan, M. H., Sandral, G. A., Kidd, D. R., Lambers, H. and Simpson, R. J. (2017). Variation in root traits associated with nutrient foraging among temperate pasture legumes and grasses. *Grass Forage Sci.* 72(1), 93–103.

Yao, J. and Barber, S. A. (1986). Effect of one phosphorus rate placed in different soil volumes on P uptake and growth of wheat. *Commun. Soil Sci. Plant Anal.* 17(8), 819–827.

Ye, D., Zhang, X., Li, T., Xu, J. and Chen, G. (2018). Phosphorus-acquisition characteristics and rhizosphere properties of wild barley in relation to genotypic differences as dependent on soil phosphorus availability. *Plant Soil* 423(1–2), 503–516.

Zhang, C., Simpson, R. J., Kim, C. M., Warthmann, N., Delhaize, E., Dolan, L., Byrne, M. E., Wu, Y. and Ryan, P. R. (2018). Do longer root hairs improve phosphorus uptake? Testing the hypothesis with transgenic *Brachypodium distachyon* lines overexpressing endogenous RSL genes. *New Phytol.* 217(4), 1654–1666.

Zhao, J., Fu, J., Liao, H., He, Y., Nian, H., Hu, Y., Qiu, L., Dong, Y. and Yan, X. (2004). Characterization of root architecture in an applied core collection for phosphorus efficiency of soybean germplasm. *Chin. Sci. Bull.* 49(15), 1611–1620.

Zhu, J. and Lynch, J. P. (2004). The contribution of lateral rooting to phosphorus acquisition efficiency in maize (*Zea mays*) seedlings. *Funct. Plant Biol.* 31(10), 949–958.

Zhu, J., Zhang, C. and Lynch, J. P. (2010). The utility of phenotypic plasticity of root hair length for phosphorus acquisition. *Funct. Plant Biol.* 37(4), 313–322.

Chapter 4

Advances in understanding the environmental effects of phosphorus fertilization

Andrew N. Sharpley, University of Arkansas, USA

1 Introduction

1.1 Agricultural phosphorus and water quality

Since the late 1960s, point sources of water quality impairment have been reduced due to their ease of identification. However, water quality problems remain, and as further point-source control becomes less cost-effective, attention is being directed towards the role of agricultural nonpoint sources in water quality degradation worldwide (Li et al., 2015; Sun et al., 2012; Withers and Jarvie, 2008). For instance, in 1996, over half of surveyed waters in the United States were nutrient impaired (Dubrovsky et al., 2010).

Nearly 20 years on, continuing water quality impairment in the United States has led to major nutrient sustainability and land conservation initiatives to reduce nutrient loss. These initiatives include the Chesapeake Bay Watershed (Kovzelove et al., 2010; U.S. Environmental Protection Agency, 2010), Lake Erie Basin (Jarvie et al., 2017; Michalak et al., 2013; Smith et al., 2015a, 2018) and Mississippi River Basin (National Research Council, 2008).

Phosphorus (P) inputs to freshwaters can accelerate eutrophication (Carpenter et al., 1998). Although nitrogen (N) and carbon (C) are also essential to the growth of aquatic biota, most attention has focussed on P because of the difficulty in controlling exchange of N and C between the atmosphere and

http://dx.doi.org/10.19103/AS.2019.0062.07

water, and fixation of atmospheric N by some blue-green algae (Schindler et al., 2008).

Thus, mitigating freshwater eutrophication is largely dependent on controlling P inputs (Carpenter, 2008). However, for water bodies with naturally high salt content, as in estuaries, reduction of N and, to a certain extent, P inputs are generally required to limit aquatic productivity (Howarth et al., 2000).

In the last 15 years, efforts to reduce or control eutrophication and to lessen further impairment have transitioned from dealing with the effects of eutrophication to identifying and targeting the sources of nutrients in a watershed for treatment. As a result, targeted management strategies are now in place in most impaired waters of the United States, as for example in the New York City drinking water supply watersheds, Chesapeake Bay Watershed, Florida's inland and coastal waters, Lake Erie Basin and Mississippi River Basin (Sharpley et al., 2003; U.S. Department of Agriculture and Environmental Protection Agency, 1999). This was an important shift, which led to strategically targeting management strategies and conservation practices to critical areas of the landscape, and is key to minimizing the impact of agricultural P-related water quality issues (Kleinman et al., 2017; Sharpley et al., 2017).

1.2 Phosphorus and food, energy and water security

Any discussion of nutrient-related water quality impairment has to be balanced with the dramatic increase in agricultural production and P-use efficiency with the development and use of mineral phosphate fertilizers in the last 50 years. This is now coupled with advances in crop breeding, soil and crop tissue testing, variable rate application of fertilizer, precise fertilizer applications and precision conservation in the last 20 years.

Phosphorus is an essential element for crop and livestock production. Over the past 50 years, global fertilizer P use has increased 350% and food production more than doubled (Khan et al., 2009). Along with this, global flows of P have increased fourfold (Childers et al., 2011; Haygarth et al., 2014), with distinct areas of grain and animal production functioning in geographically disparate, yet cost-efficient systems. In addition, 80% of the P mined from phosphate rock does not make it to food consumed by the global population (Neset and Cordell, 2012), with only 10% in human wastage recycled back to agricultural lands (Elser and Bennett, 2011).

These inefficiencies of P use are of increasing concern for three reasons. Firstly, unlike N, which is a renewable atmospheric resource, phosphate rock is a finite, non-renewable resource and with economically extractable supplies geographically limited (Jasinksi, 2015). Recent analyses suggest that at the current rate of use, rock P reserves would be depleted in around 300 years

(Scholz and Wellmer, 2013). Given these supplies of rock P are limited to a few regions of the world, rock P and fertilizer P production has the potential to be an important factor determining food security (Fig. 1).

Secondly, mandates to expand biofuel production worldwide to increase future energy security has added pressure on P fertilization of biofuel feedstocks (Hein and Leemans, 2012). Biofuel feedstocks such as sugarcane, wheat, corn and sugar beet for bioethanol production, and rapeseed, soybean and palm oil for biodiesel production, now compete for land, water and P use with food production. These pressures can increase the risk of P loss to surface waters (i.e. affecting water security) and increased grain prices (i.e. affecting food security) (Robertson et al., 2008; Tilman et al., 2009).

Finally, the increasing incidence and severity of surface water eutrophication and associated harmful algal toxic blooms worldwide has recently started to impact urban populations (Smith et al., 2018; Stumpf et al., 2016; Wines, 2014). Clearly, P is a key element to the stability and security of the food, energy and water (Jarvie et al., 2015; Fig. 1).

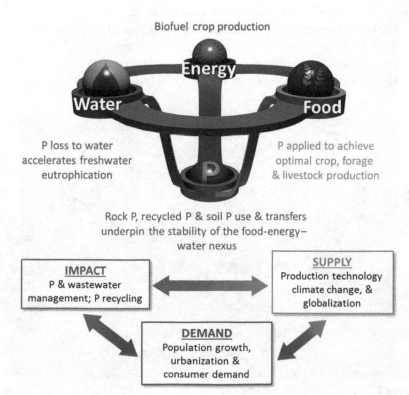

Figure 1 The role of phosphorus in nexus of food, energy and water. Source: adapted from Jarvie et al. (2015).

2 Cycling, fate and transport of phosphorus in agriculture

2.1 Phosphorus cycles as agriculture evolves

Prior to the Second World War, farming communities tended to be self-sufficient, in that enough feed was produced locally to meet animal requirements and the manure nutrients could be effectively recycled to local soils to meet crop nutrient needs (Aldrick, 1980). As a result, a sustainable food chain tended to exist.

With the advent of new technologies, mechanization, increased chemical use and government incentives, agricultural production has become focussed on less agricultural land and on fewer, but larger, farms (Evans et al., 1996). Between 1950 and 1990, U.S. farmland decreased from 500 million ha to 400 million ha (20%) and the number of farms from 5.6 million to 2.1 million (63%), while average farm size increased from 90 ha to 200 ha (120%).

Because of this regional intensification, modern farming systems have become fragmented with increasing separation of crop and livestock production, even across regional boundaries. Similar trends in historical changes have been noted in Europe (Withers et al., 2002).

Because of the spatial separation of crop and animal production systems, P fertilizers are imported to areas of grain production (Fig. 2). The grain and harvested P are then transported to areas of animal production, where inefficient animal utilization of nutrients in feed (<30% utilized) are excreted

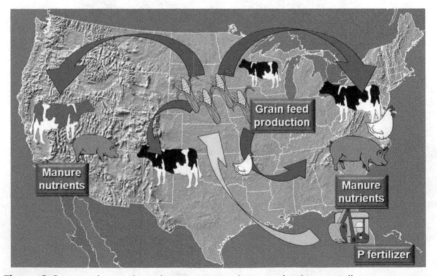

Figure 2 Grain and animal production systems have evolved in spatially separate areas leading to localized accumulations of nutrients in manure.

as manure. This has led to a large-scale, one-way transfer of nutrients from grain- to animal-producing areas and dramatically broadened the emphasis of nutrient management strategies from field and watershed levels to national scales (Fig. 2; Lander et al., 1998; Lanyon, 2005).

As a result, the potential for P surpluses in regions with a concentration of confined livestock-feeding operations can be much greater than in cropping systems, where nutrient inputs become dominated by livestock manure rather than fertilizer purchased explicitly for application to crops. As the intensity of animal production within a watershed increases, more P must be recycled, the P farm surplus (input minus output) becomes greater, soil P levels increase and the overall risk of P run-off tends to increase (Daverede et al., 2003; Haygarth et al., 1998; Pote et al., 1996; Torbert et al., 2002; Withers et al., 2002).

2.2 The fate of phosphorus in agriculture

The fate of P in agriculture is depicted in Fig. 3. Typically, less than a third of feed P is utilized by animals, with the remainder excreted in manure and applied to land for crop use. Several soil and crop factors, including soil P sorption capacity; crop type; P application type, method and rate; and land management, influence plant P uptake (Pierzynski and Logan, 1993; Sims and Sharpley, 2005).

Phosphorus losses are often agronomically small (generally <2 kg P ha^{-1}), representing a minor proportion of applied fertilizer or manure P (generally $<5\%$). Phosphorus uptake and harvest removal by crops ranges from 10% to 40% of applied P (Fig. 3), due to the rapid and only slowly reversible sorption of P to Al, Fe and Ca compounds in soil. Nevertheless, losses of <1 kg P ha^{-1} have been shown to accelerate eutrophication of receiving waters (Smith et al., 2015b).

Phosphorus loss by leaching is generally greater in sandy, organic or peaty soils – those with low P adsorption capacities – and in soils with substantial preferential flow pathways (Bengtson et al., 1988; Sharpley and Syers, 1979; Sims et al., 1998) (Fig. 3).

Phosphorus loss from tile-drained fine-textured soils receiving manure is generally low, although the increase in the number and intensity of tiled fields can increase the source area contributing nutrients directly to a stream and bypassing the mediating soil matrix. Soils that allow substantial subsurface export of dissolved P are common in parts of the northeast Coastal Plains region of the Delmarva Peninsula of Delaware and Maryland, United States, and are important to be considered in minimizing coastal water eutrophication in these regions (Sims et al., 1998).

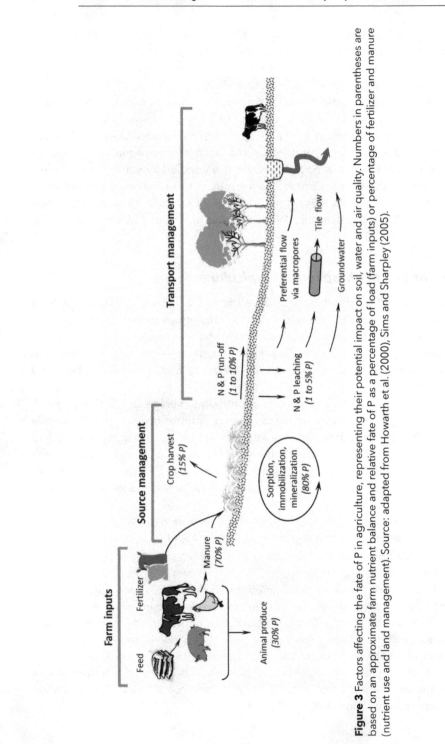

Figure 3 Factors affecting the fate of P in agriculture, representing their potential impact on soil, water and air quality. Numbers in parentheses are based on an approximate farm nutrient balance and relative fate of P as a percentage of load (farm inputs) or percentage of fertilizer and manure (nutrient use and land management). Source: adapted from Howarth et al. (2000), Sims and Sharpley (2005).

2.3 The transport of phosphorus in agricultural run-off

In the absence of surface water quality standards oriented toward minimizing eutrophication in the early 1990s, and without research data, several states in the United States recommended threshold soil test P levels that are perceived to limit eutrophic run-off. However, care must be taken on how soil test P results are interpreted for environmental purposes.

Interpretations given on soil test reports (e.g. low, medium, optimum and high) were established based on the expected response of a crop to P. Some people simply extended the levels used for interpretation of crop response to say that if soil test P was above the level where a crop response is expected, then it is in excess of crop needs and, therefore, is potentially enriching run-off with P (Fig. 4).

However, two factors are of critical importance to the debate on how to use soil test P in environmental risk assessment. Firstly, the gap between crop and environmental soil P thresholds reflects the difference in soil P removed by an acid or base extractant (i.e. Mehlich, Bray, Olsen agronomic tests) and by less invasive water (i.e. simulating extraction of soil P by run-off water), which is soil specific (Fig. 4). Secondly, soil P is only one of several factors influencing the potential for P loss; therefore, soil test P should not be used as the sole criterion on which to base P management planning.

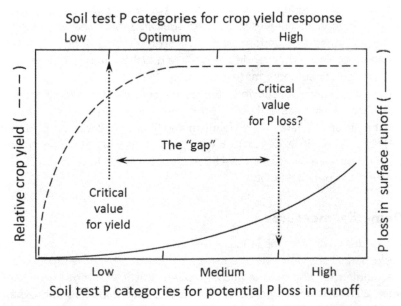

Figure 4 As soil P increases so does crop yield and the potential for P loss in surface run-off. The interval between the critical soil P value for yield and run-off P will be important for P management.

Several studies have found a change or break point in the relationship between soil test P and the concentration of P in surface run-off and subsurface flow of leached water (Table 1). One of the first to report this was Heckrath et al. (1995), who found that soil test P (as Olsen P) >60 mg kg^{-1} in the plough layer of a silt loam caused the dissolved P concentration of tile drainage water to increase significantly (0.15-2.75 mg L^{-1}). They postulated that this high concentration (>60 mg Olsen P ha^{-1}), which is well above that needed by major crops for optimum yield (about 20 mg kg^{-1}), is a critical change point above which the potential for P movement in land drains greatly increases. These and other change points are listed in Table 1.

Another method used to determine environmental soil P thresholds is estimation of the degree of P sorption saturation (DPS), which is based on the premise that the saturation of P sorbing sites in soil determines P release (intensity factor), as well as the level of soil P (capacity factor) (Breeuwsma and Silva, 1992; Kleinman and Sharpley, 2002). For example, soils of similar soil test P may have differing capacities to release P to run-off because P would be bound more tightly to clay than sandy soils (Sharpley and Tunney, 2000).

Phosphorus sorption saturation can also represent the capacity of soil to sequester further P addition and thereby limit its potential to enrich run-off P (Lookman et al., 1996; Schoumans et al., 2015). The addition of P to soil with a high DPS will enrich run-off P more than if P was added to soil with a low P sorption saturation, independent of soil test P (Leinweber et al., 1997; Sharpley, 1995a). Traditional techniques to estimate soil DPS have relied upon methods that are not commonly performed by soil testing laboratories, such as acid ammonium oxalate extraction in the dark (e.g. Schoumans and Breeuwsma, 1997) and P sorption isotherms (e.g. Sharpley, 1995b).

Recent research has shown DPS in acidic soils can be reliably estimated from Mehlich-3 extractable Al and Fe (primary components of P sorption) and P (Beauchemin and Simard, 1999; Kleinman and Sharpley, 2002; Nair and Harris, 2014). Change points in DPS, above which the concentration of P in run-off or release to soil water increases, have been found to range from 15% to 56% for several studies detailed in Table 1.

3 Remedial measures

Great strides have been made in conservation practice adoption over the last decade, particularly with watershed-specific initiatives that have provided cost-share funding for appropriate and approved practices (Kröger et al., 2012; Miltner, 2015). However, there has been a less than expected improvement in water quality despite these conservation efforts due in part to lag times associated with hydrologic flowpaths, watershed response times and the

Table 1 Change point values reported for the relationship between soil test P estimates (x) (i.e. plant available P and run-off or leachate P estimates (y))

Reference	Location	Number of observations	Soil P estimate (x)	P loss estimate (y)	Change point (mg kg⁻¹)	Regression slope	
						Before change point	After change point
Soil test P estimate (mg kg⁻¹)							
Bond et al. (2006)	North Carolina	25	Mehlich-3	Water-soluble soil P	115	0.02	0.20
Heckrath et al. (1995)	England	~33	Olsen P	Dissolved leachate P	56	–	–
Jordan et al. (2003)	N. Ireland	42	Olsen P	Dissolved run-off P	22	0.001	0.048
McDowell and Sharpley (2001)	England	43	Olsen P	Dissolved leachate P	35	–	–
	Pennsylvania	75	Mehlich-3	Dissolved run-off P	185	–	–
				Dissolved leachate P	193	–	–
Maguire and Sims (2002a)	Delmarva Peninsula, Delaware and Maryland	105	Water	Leachate dissolved P	8.6	0.025	0.12
			0.01 M CaCl$_2$		8.6	0.034	0.25
			Mehlich-3		181	0.0003	0.0124
Sims et al. (2002)	Delaware	120	Mehlich-3	Dissolved leachate P	235	0.0023	0.0147
Degree of soil P sorption saturation (%)							
Butler and Coale (2005)	Beltsville, Maryland	40	Oxalate	Water-soluble soil P	34	0.11	0.61
	Poplar Hill, Maryland	40	Oxalate	Water-soluble soil P	25	0.04	0.80
	Queenstown, Maryland	40	Oxalate	Water-soluble soil P	30	0.07	1.10
	Upper Marlboro, Maryland	40	Oxalate	Water-soluble soil P	28	0.10	0.79
Casson et al. (2006)	Alberta	47	Mehlich-3	Water-soluble soil P	3–44	–	–
Hooda et al. (2000)	England	320	Oxalate	Water soluble soil P	10	–	–
Maguire and Sims (2002b)	Delaware	105	Oxalate	Leachate dissolved P	56	0.0026	0.108
Nair and Harris (2004)	Florida	69	Mehlich-3	Water soluble soil P	16	0.060	0.201
Nelson et al. (2005)	North Carolina	60	Oxalate	Water soluble soil P	45	0.001	0.140
Sims et al. (2002)	Delaware	120	Mehlich-3	Dissolved run-off P	0.13	0.024	4.33
				Dissolved leachate P	0.2	0.0098	28.44

legacies of past management (Jarvie et al., 2013; Kronvang et al., 2005; Meals et al., 2010; Murphy et al., 2015; Sharpley et al., 2013).

Many conservation practices can be implemented over a wide range of scales to minimize the loss of P from agriculture to surface and ground waters (Table 2). These are commonly grouped into measures that seek to reduce the input of P onto farms and bring them into closer balance with outputs in produce; manage on-farm nutrient sources through appropriate rate, timing and method of P application; and measuring the potential for P transport to surface and ground waters (Table 2). These measures are also depicted in Fig. 5.

3.1 Nutrient management

Careful nutrient management planning on a field-by-field and whole-farm basis is a major component of any remedial action plan to minimize the risk of nutrient loss from agricultural lands. This basically follows the '4R' nutrient management approach espoused by the International Fertiliser Association (2009) and International Plant Nutrition Institute (2014), which is adding P in the right form, right rate to match crop needs, in the right place and at the right time (Fig. 5).

New programmes, such as the '4R Plus' programme in Iowa (https://www.4rplus.org/), where 'Plus' refers to conservation practices that can boost production, increase soil resiliency, reduce erosion and run-off and improve water quality in addition to earlier mentioned 4R nutrient management (Iowa Nutrient Reduction Strategy, 2012). The 4R Plus programme encourages producers to adopt practices like these, which will both increase nutrient-use efficiency and decrease the environmental impact of nutrients.

As producers strive to attain optimal timing for nutrient application, they may inadvertently place additional stresses and challenges on companies responsible for fertilizer logistics, storage and transport. Furthermore, additional field operations may need to be added as optimal timing for P applications are often different for that over other nutrients, therefore nutrients that were usually co-applied may require their own application operation.

3.1.1 Right Form of nutrient

Fertilizer nutrients can be formulated to match crop needs; however, manures have more P than N compared to crop needs. For instance, the ratio of N:P in manure (2–4:1) is two to four times lower than that taken up by major grain and hay crops (8:1). As a result, application of manure to meet crop N needs results in the application of three to four times more P than the crop needs annually.

Table 2 Conservation practices to minimize the loss of P from agriculture

Practice	Description	Impact on P loss[a]
Farm inputs		
Crop genotypes	Low phytic acid corn reduces P in manure	Decrease
Feed additives	Enzymes increase P utilization by animals	Decrease
Feed supplements	Match animals nutritional P requirements	Decrease
Source management		
Crop requirements	P applications based on crop needs	Decrease
Soil P testing	Nutrient applications based on soil P availability	Decrease
Cover crops/residues	If removed offsite can reduce residual soil P	Decrease TP Decrease DP
Method of application	Incorporated, banded or injected in soil	Decrease
Rate of application	Match crop needs	Decrease
Source of application	Sources can differ in their P availability	Decrease
Timing of application	Avoid application to frozen ground Apply during season with low run-off probability	Decrease
Transport management		
Conservation tillage	Reduced tillage and no-till increases infiltration and reduces soil erosion	Decrease TP Increase DP
Strip cropping, contour tillage, terraces	Reduced transport of sediment-bound nutrients	Decrease TP Neutral DP
Conservation cover	Permanent vegetative cover increases soil infiltration and water-holding capacity	Decrease
Invert stratified soils	Redistribution of surface P through profile by ploughing, spading, with and without inclusion plates, and deep ripping	Decrease
Buffer, riparian, wetland areas, grassed waterways	Remove sediment-bound P, enhance denitrification	Decrease TP Neutral DP
Critical source area treatment	Target sources of P in a watershed for remediation	Decrease

[a] TP is total P and DP is dissolved P.

Repeatedly applying manure at rates to provide sufficient N increase soil P levels and the risk of P run-off. On the other hand, application of manure to meet crop P needs would apply insufficient N, putting an economic burden on farmers to purchase costly mineral N.

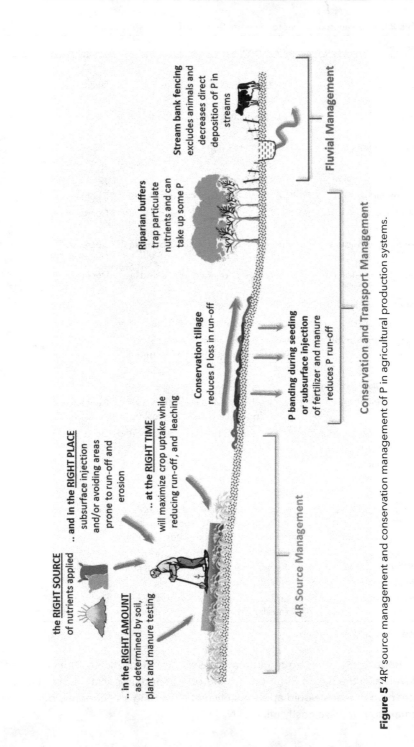

Figure 5 '4R' source management and conservation management of P in agricultural production systems.

3.1.2 Right rate via soil testing

Fertilizer P rates are usually established by crop need and modified by the amount already in the soil, as determined by established soil test P methods (Cox, 1994). In the case of commercial fertilizer P, applications can easily be tailored to match crop needs and minimize excessive soil P accumulation because an economic disincentive exists to avoid applying excess fertilizer P.

For instance, Dodd and Mallarino (2005) showed that annual fertilizer P rates of about 15 kg P ha^{-1} maintained near-optimum soil test P levels (16–20 mg kg^{-1} as Bray-1 P) and corn [*Zea mays* (L)]-soybean [*Glycine max* (L.) Merr.] yields for Mollisols in Iowa. Until recently, however, manure applications have been made to meet the N needs of the crop, which has resulted in a build-up in soil test P above levels needed for optimum crop yields, and increased the risk of P loss in run-off.

3.1.3 Right place

Because of the relative immobility of P in the soil profile, placement of fertilizer P generally is more critical for plant availability than in the case of fertilizer N. Long-term studies in the northern Great Plains show that high rates of broadcast P (90 kg P ha^{-1}) can have long-term effects (17 years) on soil test P, wheat yields (Bailey et al., 1977; Halvorson and Black, 1985) and profitability (Jose, 1981; Halvorson et al., 1986). Several studies show a greater yield response to surface or subsurface band application of fertilizer P at low rates, compared to broadcasting or mixing (Alston, 1980; Bailey and Grant, 1989; Lamond, 1987; Yost et al., 1981). In fact, Welch et al. (1966) observed greater P uptake and yield for corn with a combined banded (50%) and broadcast (50%) application (total of 40 kg ha^{-1}).

The incorporation of manure into the soil profile, either by tillage or subsurface placement, reduces the potential for P loss in surface run-off. Rapid incorporation of manure also reduces ammonia volatilization and potential loss in run-off, as well as improving the N:P ratio for crop growth. For example, Mueller et al. (1984) showed that incorporation of dairy manure by chisel ploughing reduced total P loss in run-off from corn 20-fold, compared to no-till areas receiving surface applications. In fact, P loss in run-off declined because of a lower concentration of P at the soil surface and a reduction in run-off with incorporation of manure (Mueller et al., 1984; Pote et al., 1996).

3.1.4 Right time of application

Many studies show the loss of P in run-off relates directly to the rate and frequency of applied fertilizer P (Sharpley et al., 2007; Sims and Kleinman, 2005). Similarly, manure application and timing relative to rainfall influence the concentration of P in run-off (Dampney et al., 2000; Sims and Kleinman, 2005). Several studies show, for example, reductions in P losses with an increase in the length of time between manure application and surface run-off (Djodjic et al., 2000; Edwards and Daniel, 1993; Sharpley, 1997; Westerman et al., 1983). These reductions can be attributed to the reaction of added P with soil and dilution of applied P by infiltrating water from rainfall that did not cause surface run-off.

Rainfall intensity and duration, as well as when rainfall occurs relative to when manure is applied, all are factors that influence the concentration and loss of manure P in run-off. The relationship between potential loss and application rate, however, is critical to establishing conservation strategies. Even though soil P clearly is important in determining P loss in surface run-off, the rate and frequency of applying P to soil can override soil P in determining P loss (Sharpley et al., 2007).

Also evident are the long-lasting effects of applied manure on increased concentrations of P in surface run-off. For instance, Pierson et al. (2001) found that a poultry litter application tailored to meet pasture N demands elevated surface run-off P for up to 19 months after application.

3.2 Transport management

Transport management refers to efforts to control the movement of P from soils to sensitive locations such as bodies of freshwater. Phosphorus loss via surface run-off and erosion may be reduced by conservation tillage and crop residue management, buffer strips, riparian zones, terracing, contour tillage, cover crops and impoundments or ponds (Table 2 and Fig. 5). These practices tend to reduce rainfall impact on the soil surface, reduce run-off volume and velocity, and increase soil resistance to erosion. However, none of these measures should be relied on as the sole or primary practice to reduce P run-off.

3.2.1 Run-off potential

The potential for run-off from a given site is important in determining the loss of P and is thus a critical component of nutrient management strategies. For instance, Pionke et al. (1999, 2000) generalized that about 90% of annual P export (0.63 kg ha^{-1} year^{-1}) from an agricultural watershed in the Chesapeake Bay Basin occurs during storm flow because of run-off from saturated areas extending no more than 60 m from the stream channel. In contrast, only 20% of N (6 kg ha^{-1} year^{-1}) is exported annually in storm run-off.

Distance from where run-off is generated to a stream channel influences P loss and thus must be a nutrient management consideration (Gburek and Sharpley, 1998). Run-off, and nutrients carried by it, can be reduced or even intercepted by infiltration and deposition prior to reaching a stream channel. Generally, the further a field is from a stream channel, the lower the potential for run-off to contribute nutrients to the stream. Many states in the United States, therefore, have adopted the premise of implementing more restrictive nutrient management strategies, particularly for P, on fields close to streams (Sharpley et al., 2003).

3.2.2 Subsurface drainage potential

Within the 4R framework, soil testing (Sharpley et al., 2001; Maguire and Sims, 2002a; Duncan et al., 2017), subsurface placement (Williams et al., 2016) and timing of application (King et al., 2018; Schroeder et al., 2004; Smith et al., 2007) have all been shown to reduce P loss in tile drainage. Historical recommendations for P fertility focussed on a build-and-maintain philosophy that has led to large P stores in many fields (Jarvie et al., 2013; Withers et al., 2014). Soil testing at least once in the rotation and adhering to those recommendations creates a change in philosophy from 'build and maintain' (i.e. 'feed the soil') to a 'feed the crop' mentality, which in time, should reduce the chronic losses of P (King et al., 2018; Withers et al., 2014).

3.2.3 Erosion potential

A sequence of changing tillage practices in several watersheds in Oklahoma enabled comparison of surface and ground water impacts associated with native grass, conventionally tilled wheat and no-till wheat (Sharpley and Smith, 1994). In the late 1970s, conversion of native grass to conventionally tilled wheat increased soil loss significantly (Table 3). From 1979 to 1990, fertilizer P applications associated with conventionally tilled wheat also contributed to increased losses (0.55 kg P ha^{-1} year^{-1}) in surface run-off compared to native grass (Table 3).

Following conversion of two watersheds to no-till wheat in 1984, erosion was greatly reduced (11 300-625 kg^{-1} soil ha^{-1} year^{-1}; Table 3), due to the protective vegetative cover of no-till wheat residues. Relatedly, total P loss in run-off from no-till wheat (1.4 kg ha^{-1} year^{-1}) was appreciably less than that from conventionally tilled wheat (4.9 kg ha^{-1} year^{-1}; Table 3). However, the loss of dissolved P in run-off increased following the conversion to no-till wheat (0.3-0.7 kg ha^{-1} year^{-1}; Table 3). In other words, the proportion of P run-off that was immediately bioavailable (i.e. dissolved P form) increased from 5% for conventionally tilled wheat to 52% for no-till wheat.

Similar amounts of fertilizer P were applied to the conventionally tilled and no-till wheat watersheds (about 60 kg P ha^{-1} year^{-1}; Sharpley and Smith, 1994). The increase in dissolved P run-off can be attributed to leaching of P from crop residue and preferential transport of clay-sized particles under no-till compared

Table 3 Annual sediment and P loss in run-off as a function of tillage management of watersheds at El Reno and Woodward, Oklahoma

		Loss in run-off		
		Sediment	Dissolved P	Total P
Study period	Management		(kg ha^{-1} year^{-1})	
1977–78	Native grass	103	0.03	0.08
1979–83	Native grass	38	0.05	0.09
	Conventional-till wheat	575	0.12	0.64
1984–90	Native grass	34	0.07	0.15
	Conventional-till wheat	11 300	0.26	4.87
	No-till wheat	625	0.74	1.43

Source: adapted from Sharpley and Smith (1994).

to conventional tillage. In addition, such practices as strip and contour cropping reduce soil erosion potential and enhance infiltration, thereby minimizing an important P loss pathway in surface run-off and erosion.

3.2.4 Permanent vegetation

Keeping land in permanent cover, such as grass or cover crops, reduces the risk of run-off and erosion, increases infiltration and thereby minimizes the loss of P. Cover crops protect the soil surface from raindrop impact, improve infiltration relative to bare soil and trap eroded soil particles (Sharpley and Smith, 1991). Where dissolved P transport is the primary concern, cover crops may reduce run-off and, consequently, run-off P load. However, cover crops are unlikely to affect dissolved P concentration in run-off. Kleinman et al. (2005) found that cover crops reduced total P concentration in springtime run-off to 36% of the dissolved P in run-off from conventional corn. However, dissolved P concentration was not significantly different between cover crops and conventional corn because dissolved P was controlled by soil P content rather than by soil erosion.

Grassed waterways are designed to trap sediment and reduce channel erosion. In some cases, waterways are installed as cross-slope diversions to intercept run-off and reduce effective slope length. Chow et al. (1999) estimated that installation of grassed waterways and terraces in combination reduced annual soil erosion 20-fold in a New Brunswick, Canada, potato field.

3.2.5 Riparian/buffer areas

Healthy riparian areas can reduce P export, increase wildlife numbers and diversity and improve aquatic habitat. In addition to acting as physical buffers to sediment-bound nutrients, plant uptake captures P, resulting in short-term and long-term accumulations of P in biomass (Hoffmann et al., 2009; Uusi-Kämppä et al., 2000).

The effectiveness of riparian buffers as a nutrient management practice can vary significantly. For instance, the route and depth of subsurface water flow paths through riparian areas can influence nutrient retention. Riparian buffers are most efficient when sheet flow occurs, rather than channelized flow that often bypasses some of the retention mechanisms. Those areas must be managed carefully to realize their full retention and filtration capabilities.

3.2.6 Critical source area management

Transport of P from agricultural watersheds depends to a large extent upon the coincidence of source (soil, crop and management) and transport factors (run-off, erosion and proximity to watercourse or body). Source factors relate to watershed areas with high potential to contribute to P export. For P, source areas often are spatially confined and limited in extent, generally reflecting soil P status and fertilizer and manure P inputs (Gburek and Sharpley, 1998; Pionke et al., 2000).

Phosphorus transport generally occurs from hydrologically active areas in a watershed where surface run-off contributing to stream flow is coincident with areas of high soil P (Gburek and Sharpley, 1998; Gburek et al., 1996). For example, Pionke et al. (1996) reported that 90% of annual P exported ($0.63 \ kg^{-1} \ ha^{-1} \ year^{-1}$) from a south central Pennsylvanian watershed occurred during storm flow as surface run-off on saturated areas extending no more than 60 m from the stream channel and which itself accounted for only 10% of annual run-off. In contrast, 20% of nitrate export was associated with storm events; the remaining 80% ($6 \ kg^{-1} \ ha^{-1} \ year^{-1}$) was exported in baseflow that came from a larger recharge area of the watershed.

Even in regions where subsurface flow pathways dominate, areas contributing P to drainage water appear to be restricted to soils with high soil P saturation and hydrologic connectivity to the drainage network. For example, Schoumans and Breeuwsma (1997) found that soils with high P saturation contributed only 40% of total P load, while another 40% came from areas where the soils had only moderate P saturation but some degree of hydrological connectivity with the drainage network.

3.3 Conservation practice implementation and response at a watershed scale

Conservation practices have been shown to decrease edge-of-field losses of P, but does this relate to an improvement in water quality at a watershed scale? Table 4 summarizes the results from several long-term monitoring

Table 4 Reduction in total, dissolved and particulate P loads found in the U.S. watershed monitoring schemes following the implementation of a range of conservation practices

Study	Location	Receiving watershed	Major land use	CPs implemented (main target)[a]	Study type (duration)		% reduction		
							TP	PP	DP
Bishop et al. (2005)	Cannonsville Reservoir Basin (NY)	Cannonsville Reservoir	Dairy pasture and crop	Manure management (DP) Rotational grazing (PP) Drainage management (DP) Crop management (PP)	Paired watershed study. Pre- vs post-implementation monitoring (2 years pre-, 5 years post-)		n.d.	29	43
Brannan et al. (2000)	Owl Run Watershed (VI)	Chesapeake Bay	Dairy pasture and crop	Manure management (DP) Stream fencing (DP+PP) Cover crops (PP) Field strip cropping (PP) Grass waterways (PP)	Pre- vs. post-implementation monitoring (3 years pre-, 7 years post-)	Watershed outlet	54	66	23
						Dairy dominated sub-watershed	25	35	4
						Crop dominated sub-watershed	36	70	−117
Inandar et al. (2001)	Nomini Creek (VI)	Chesapeake Bay	Crop	Nutrient management (DP) No-till (PP) Filter strips (PP) Stabilization structures (PP)	Pre- vs. post-implementation monitoring (3 years pre-, 7 years post-)		4–21	30–41	– 61–86
Richards et al. (2002)	Sandusky River (Ohio) Maumee River (Ohio)	Lake Erie	Crop Crop	Nutrient management (DP) Conservation tillage (PP)	River monitoring 1975–95 (20 years)	Sandusky River	46	n.d.	88
						Maumee River	42	n.d.	85
Daloğlu et al. (2012)	Sandusky River (Ohio)	Lake Erie			River monitoring 1995–2008 (13 years)	Sandusky River	n.d.	n.d.	70
Michalak et al. (2013)	Maumee River (Ohio)	Lake Erie				Maumee River	−58	−31	−218

[a] Based on expected effectiveness for each P form.

n.d. = not determined, TP = total P, PP = particulate P and DP = dissolved P.

studies designed to assess the impact of implementing a range of different conservation practices on P transport to sensitive water bodies. In the majority of the studies, the implemented conservation practices targeted particulate P loss, although nutrient management was common to all but one.

Significant reductions in total P loading were observed during the first 20 years of monitoring in Lake Erie Basin. However, this appears to be mostly due to a reduction in particulate P loss by adoption of cover crops, vegetated buffer strips and no-till cropping.

Reductions in dissolved P loading in Cannonsville Reservoir Basin, New York, were attributed mainly to improved nutrient management (Bishop et al., 2005). Infrastructure improvements including construction of a manure storage lagoon and improvements in roadways allowed more strategic application of manure in terms of timing, rate and more even distribution to fields further from the watercourse.

Results from long-term watershed monitoring clearly show that different conservation practices are variably effective for different forms of P, and reducing dissolved P loss appear to be especially challenging. The main strategy effective at reducing off-farm dissolved P loss is source or nutrient management (Table 2 and Fig. 3); however, this appears to have limited impact at the watershed scale (Table 4).

In contrast, practices that targeted sediment and therefore particulate P loss, namely conservation tillage, grassed waterways and terraces, were most widely adopted, regardless of whether dissolved or particulate P was of most concern (Osmond et al., 2012). Farmer surveys have indicated that they are more likely to address water quality issues for contaminants they can see (i.e. sediments) than invisible dissolved P (Reimer et al., 2012), explaining the preference for erosion control measures.

4 Legacies of past management

Prior land management activities can lead to a long-term legacy of P in watersheds (Fig. 6). Legacy P refers to P stored in surface soils, ditches, riparian zones, wetlands, stream and lake sediments from prior land and nutrient management (Table 5; Meals et al., 2010; Sharpley et al., 2013). The stored P can be subsequently re-released as the P storage capacity gradually becomes saturated, or after a change in land use or land management.

Lag times associated with release of P from legacy P stores may help explain difficulties in detecting water quality improvements at a watershed scale. For example, where soil test P levels have risen to more than ten times crop sufficiency levels, it can take a decade or more to 'draw down' soil P reserves to levels where dissolved P in run-off is substantially reduced (Cox et al., 1981; Shulte et al., 2010).

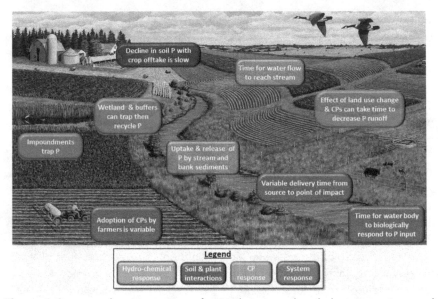

Figure 6 Conceptual representation of natural, managed, soil-plant interactions and conservation practice processes influencing the lag time for system response.

In fluvial systems, the release of legacy P from deposited sediments can be influenced by the oxygen status of overlying waters, where reducing conditions favours the release of sediment-bound P. In addition, the beneficial effects of conservation practices on lower P inputs to stream and rivers decreases P concentrations that trigger the release of P from sediments, masking downstream benefits of conservation practices on P fluxes (Dodd and Sharpley, 2015, 2016). Finally, P enriched sediments are re-suspended during high stream flows.

By developing watershed-scale monitoring that identifies local-scale improvements and associated time lags in water quality as they occur, watershed planners can start to better understand and plan mitigation strategies. For example, nutrient fluxes into Western Lake Erie Watershed, land management, lake water quality and outbreaks of harmful algal blooms have been monitored for the last 50 years. This provides an excellent example of how well-meaning conservation strategies can result in intended and unintended consequences on nutrient fate and transport (Jarvie et al., 2017; Richards et al., 2009; Smith et al., 2015a). Steady declines in P inputs from predominantly agricultural watersheds were measured between 1980 and 1995 with the adoption of conservation practices, such as nutrient management planning that reduced fertilizer and manure applications to corn and soybeans and a transition to no-till cropping.

Simply implementing conservation practices may not always provide the desired outcomes. For example, despite widespread adoption of conservation

Table 5 Example ranges in time lags for recovery generated by legacy P contributions to soils, rivers and lakes

System	Location	Parameter	Decline[a]	Time lags (years)	Decline (mg kg^{-1} year^{-1})	Reference
Soils[b]						
Thurlow loam under small grains	Montana	Olsen P	60-6	9	6.0	Campbell (1965)
Portsmouth fine sandy loam under small grains	North Carolina	Mehlich-1 P	54-26	9	3.1	Cox et al. (1981)
Haverhill clay loam under wheat	North Carolina	Olsen P	135-70	14	4.6	Cox et al. (1981)
Portsmouth fine sandy loam under corn	North Carolina	Mehlich-3 P	100-20	16-18	4.7	McCollum (1991)
Ruston fine sandy loam under bermudagrass hay	Oklahoma	Mehlich-3 P	258-192	6	11.0	Sharpley et al. (2007)
Othello silt loam under corn	Maryland	Mehlich-3 P	488-465	5	4.6	Sharpley et al. (2009)
Range of soils under arable and pasture	Ireland	Morgan's P	>8 to 5-8[c]	7-15	0.4	Shulte et al. (2010)
Carroll clay loam under wheat	Manitoba	Olsen P	222-50	8	21.5	Spratt et al. (1980)
Waskada loam under wheat	Manitoba	Olsen P	200-50	8	18.9	Spratt et al. (1980)
Ste. Rosalie clay under corn	Montreal	Mehlich-3 P	125-109	4	4.0	Zhang et al. (2004)
Nicollet-Webster loam under corn-soybean rotation	Iowa	Bray-1 P	95-8	27	3.2	Dodd and Mallarino (2005)
Rivers						
In-channel sediment and P storage and remobilization	Lowland permeable rivers (Tern, Lambourn, Frome and Piddle), UK	Fine sediment and Particulate P	N/A	<1	N/A	Collins and Walling (2007a,b), Ballantine et al. (2006)
	River Lambourn, UK	Dissolved P	30 µg L^{-1}	0.5	N/A	Jarvie et al. (2005)
	Yorkshire rivers, UK	Fine sediment, total P	N/A	<1	N/A	Owens and Walling (2002)
Watershed-scale fluvial sediment storage and remobilization	Coon Creek, WI; Powder Creek, MT; Amazon Basin	Bulk sediment	N/A	Decades to centuries	N/A	Trimble (2010)

(Continued)

Table 5 (*Continued*)

System	Location	Parameter	Decline[a]	Time lags (years)	Decline (mg kg⁻¹ year⁻¹)	Reference
Floodplain sediment and P storage and remobilization	Murder Creek, GA	Bulk sediment	N/A	Centuries to millennia	N/A	Jackson et al. (2005)
	River Swale, UK	Fine sediment, total P, and heavy metals	N/A	Decades to centuries	N/A	Walling et al. (2003)
Reactivation of legacy P by floodplain restoration	River Yare, UK	Dissolved P	N/A	Decades to centuries	N/A	Surridge et al. (2012)
	River Lobau, Germany	Total P	N/A	Decades to centuries	N/A	Schönbrunner et al. (2012)
Riparian restoration	Vermont	Total P	20% decrease	2	N/A	Meals and Hopkins (2002)
		Dissolved P	20–50% decrease	2	N/A	Meals and Hopkins (2002)
Lakes						
Lake Müggelsee, WWTP[d] upgrades, farm CPs[e], industrial source controls, P-free detergents	Germany	Total P	230–185[i]	6	N/A	Kohler et al. (2005)
Barton Broad; WWTP upgrades, diversion of P-rich inflow	UK	Total P	308[i]–95[i]	20	N/A	Phillips et al. (2005)
Little Mere; WWTP effluent diversion	UK	Total P	2350[ii] – 167[ii]	>11	N/A	Moss et al. (2005)
Shagawa Lake; WWTP upgrades	Minnesota	Total P	51–30	5	4.2	Larsen et al. (1979)
Shagawa Lake; WWTP upgrades	Minnesota	Dissolved P	21–4.5	5	3.3	Larsen et al. (1979)

[a] Soil units are mg P kg⁻¹ soil; river units are µg P L⁻¹; and lake units are [i]summer–autumn or [ii]annual mean values as µg P L⁻¹.
[b] Soil samples collected from the surface 0–15 or 0–20 cm depth and received no P during the stated period.
[c] Morgan's P > 8 mg kg⁻¹ is ranked as excess soil P status and 5–8 mg kg⁻¹ is ranked as target soil P status.
[d] WWTP is waste water treatment plant.
[e] CP is conservation practice.

in Lake Erie Watershed, increased losses of P in agricultural run-off over the last decade have resulted from complex, dynamic and yet predictable factors. These include the accumulation of P at the soil surface, fall (autumn) application of fertilizer, continued surface broadcasting of P, a focus on implementing conservation practices that reduce erosion and particulate P loss, a rapid rise in tile drainage fuelled by higher grain prices, and release and remobilization of fluvial P.

As a result, the number of fields with tile drainage that connect to ditches and streams has increased, contributing source areas of legacy P to Lake Erie (Smith et al., 2015b). The combination of these factors created a 'perfect P loss storm', which along with more intense summer rains increased P inputs to Lake Erie to record levels in 2010, culminating in the 2014 toxic bloom and water crisis in Toledo, OH.

Scientifically valid remedial strategies are not likely to be readily adopted by farmers, due to logistical, practical and cost-related limitations (Daloğlu et al., 2012; Michalak et al., 2013; Sharpley et al., 2012; Smith et al., 2018). Thus, the research community needs to work closely with the farming community to generate innovative conservation practices, stewardship and reward programmes, which will decrease nutrient run-off and which will empower farmers.

5 Conclusion and future trends

Our understanding of the fate, transport and impacts of P in agricultural systems continues to advance; several areas for consideration are proffered from the prior discussion. These include a concerted effort to recover and recycle P from waste streams, a discussion of restructuring future agricultural production systems, managing management and impact trade-offs, increasing conservation practice adoption and the transfer of science to practice.

5.1 Recovery and recycling of P from waste streams

It is clear that greater efficiencies in P use and more effective coordination of P recovery and recycling is needed at global, regional, local and even farm levels (Table 6). The value of P in manures and urban and industrial by-products needs full recognition and has to be appropriately accounted for in watershed planning strategies, which may require innovative integration of financial incentives and/or stricter regulations. At the same time, indirect or unintended consequences associated with conflicting strategies should be avoided.

Research opportunities that may help transform some of these trade-offs into synergies for improved food-energy-water security include the development of innovative cost-effective technologies and practices for manure processing

Table 6 Examples of the roles, research opportunities and technology needs regarding P in a resilient water-energy-food security nexus

Nexus connection	The role of P	Research opportunities	Technology needs
Water and food	Use of P fertilizers has increased food grain, fibre and livestock production and food security. P loss from agricultural production systems has contributed to widespread eutrophication.	• Revisit dated Land Grant soil fertility recommendations. • Identify critical source areas and management practices for P loss. • Identify and quantify legacy sources of P within watersheds. • Unified framework to target precision conservation. • Innovative methods to recycle P at farm, watershed and global scales that reduce reliance on mined P fertilizers.	• Innovative agricultural conservation practices (CPs) that help protect and enhance soil structure and fertility, minimize P loss and increase water-use efficiency, while limiting consequences of unintended and conflicting outcomes. • Cost-effective technology to recover and recycle P from manure, and wastewater (e.g. through enhanced value products) will help close the P cycle and reduce reliance on imported inorganic P fertilizers.
Water and energy	Use of P fertilizers has enabled the specialization and intensification of agricultural production, as well as growth of biofuels industry (based on grain ethanol and biodiesel), increasing energy and water demand.	• Quantify the impact of expanding agricultural production into marginal areas on soil erosion and P loss, with longer-term trade-offs for soil C, and ecosystem services, which support food production and clean water. • Revisit soil fertility recommendations for cellulosic feedstock. • Determine resilience of farming systems to reduced water availability and potential impacts on fertilizer use.	• Ensure landscape suitability and sustainability for biofuel grain so that right biofuel crops are grown in the right place to maximize yields, while minimizing soil erosion and P loss. • Enhance landscape diversity to support a greater range of biofuel crops, for example perennial cellulosic biofuel crops can help decrease P loading to surface waters and increase soil C in less productive areas.
Food and energy	P is an essential element food and biofuel energy production.	• Economic and strategic forecasting of consequences of biofuel and food crops competing for land and water resources. • Global analysis of impacts of increasing food prices with trade-offs for food security in countries reliant on food imports. • Balance production and environmental incentives for farmers to increase production and yields while embracing conservation ethics.	• Strategic analysis of co-locating concentrated animal feeding operations near biofuel processing plants to utilize waste products of biofuel production (e.g. distiller's grain) as animal feed that increases P-use efficiency, minimizing P losses. • Cost-effective technology to generate energy from P-containing waste streams, as part of a sustainable P-recycling strategy.

and production of higher value recycled products. Policies and initiatives that promote food and energy security, via agricultural intensification, must be better coordinated and financially linked with P recycling and implementation of conservation measures that address both P scarcity and abundance issues.

5.2 Restructuring production systems

Experience tells us that there needs to be a general restructuring and realignment of production systems to more closely connect crop and livestock operations, which includes a maximum threshold level of P use and a minimum level of land conservation that avoids risky practices on vulnerable landscapes. In extreme cases of highly vulnerable landscapes, certain production systems may be inherently unsustainable, regardless of the suite of conservation practices used or conservation measures adopted.

Precision conservation and nutrient management programmes can address P source realignment (e.g. rate, method and timing of applied P) and reduce P loss through transport controls (e.g. conservation tillage, contour ploughing, cover crops and riparian buffers) to achieve the required improvements in water quality and security.

Overall, inputs in fertilizer and manures should be carefully matched with crop needs over the whole watershed. Short-term remediation of P loss should focus on critical source areas of P export where high soil P, P application and zones of surface run-off and erosion coincide. However, lasting improvements in water quality can only be achieved by balancing system inputs and outputs of both P and N. Clearly, the management of agricultural nutrients involves a complex suite of options that must be customized to meet site-specific needs. Even so, the long-term impacts have been and remain difficult to quantify.

5.3 Managing trade-offs

In reducing P loss, lessons from the indirect consequences or trade-offs of conflicting strategies must be learnt and management strategies adapted. For example, no-till conservation has dramatically decreased erosion and associated P loss. However, the loss of P in dissolved, more immediately reactive form without the incorporation of applied P can negate reductions in overall total P loss, and be sufficient to stimulate algal blooms (Richards et al., 2010; Tiessen et al., 2010).

Another trade-off resulting from the cultivation of new lands fuelled by corn for bioethanol that is facilitated by tile drainage, directly connects new source areas to stream and ditches, indirectly increasing the potential for P loss (Smith et al., 2015b). Clearly, there needs to be a more effective communication and coordination among all involved in agricultural production, policy development and strategy implementation.

5.4 Encouraging conservation practice adoption

A range of voluntary and regulatory measures can be used to encourage implementation of nutrient management strategies as part of conservation programmes to protect soil and water resources. In general, the success of these measures relates to how well farmers can afford to implement new management strategies and the concomitant level of support or incentives for their adoption.

Unfortunately, less attention has been given to mechanisms or programmes that support the maintenance of implemented conservation practices. Often, maintenance costs are appreciably greater than implementation costs, particularly as farm labour. If a farmer does not receive financial support to implement a conservation practice, then it is not likely to be voluntarily adopted unless it either makes or saves the farmer money.

5.5 Knowledge transfer through outreach and education

For real and lasting changes to occur in agricultural systems, balancing production and environmental stewardship constraints, there needs to be a greater consideration of socioeconomic drivers of what, how and why some conservation practices are adopted and others are not (Kleinman et al., 2015).

There should be a greater emphasis on education, outreach and consumer-driven programmes rather than assuming that farmers will absorb the total costs associated with implementing remedial practices. Remembering that, except for decreasing off-farm import of P and increasing on-farm P-use efficiency, conservation practices are only temporary measures to minimizing P loss and the impacts on receiving waters.

Continuing educational efforts with the public and farmers regarding the importance and impact of conservation practices on environmental quality parameters is essential to reach environmental goals. In some instances, local or regional governmental controls may be necessary to enhance prompt adoption of practices that will have a positive influence on environmental outcomes.

6 Acknowledgements

Parts of the chapter are adapted from material previously published in Sharpley (2018).

7 References

Aldrick, S. R. 1980. Nitrogen in relation to food, environment and energy. University of Illinois. Agr. Expt. Station, Special Pub. 61. University of Illinois, 451pp.

Alston, A. M. 1980. Response of wheat to deep placement of nitrogen and phosphorus fertilizers on a soil high in phosphorus in the surface layer. *Aust. J. Agric. Res.* 31(1), 13–24. doi:10.1071/AR9800013.

Bailey, L. D. and Grant, C. A. 1989. Fertilizer phosphorus placement studies on calcareous and noncalcareous chernozemic soils: growth, P-uptake and yield of flax. *Commun. Soil Sci. Plant Anal.* 20(5–6), 635–54. doi:10.1080/00103628909368106.

Bailey, L. D., Spratt, E. D., Read, D. W. L., Warder, F. G. and Ferguson, W. S. 1977. Residual effects of phosphorus fertilizer: for wheat and flax grown on chernozemic soils in Manitoba. *Can. J. Soil Sci.* 57(3), 263–70. doi:10.4141/cjss77-032.

Ballantine, D., Walling, D. E. and Leeks, G. J. L. 2006. The deposition and storage of sediment-associated phosphorus on the flood plains of two lowland groundwater fed catchments. In: Rowan, J. S., Duck, R. W. and Werritty, A. (Eds), *Sediment Dynamics Hydromorphology of Fluvial Systems*. IAHS Publ. 306, pp. 496–504.

Beauchemin, S. and Simard, R. R. 1999. Soil phosphorus saturation degree: review of some indices and their suitability for P management in Quebec, Canada. *Can. J. Soil Sci.* 79(4), 615–25. doi:10.4141/S98-087.

Bengtson, R. L., Carter, C. E., Morris, H. F. and Bartkiewics, S. A. 1988. The influence of subsurface drainage practices on nitrogen and phosphorus losses in a warm, humid climate. *Trans. Am. Soc. Agric. Eng.* 31, 729–33.

Bishop, P. L., Hively, W. D., Stedinger, J. R., Rafferty, M. R., Lojpersberger, J. L. and Bloomfield, J. A. 2005. Multivariate analysis of paired watershed data to evaluate agricultural best management practice effects on stream water phosphorus. *J. Environ. Qual.* 34(3), 1087–101. doi:10.2134/jeq2004.0194.

Bond, C. R., Maguire, R. O. and Havlin, J. L. 2006. Change in soluble phosphorus in soils following fertilization is dependent on initial Mehlich-3 phosphorus. *J. Environ. Qual.* 35(5), 1818–24. doi:10.2134/jeq2005.0404.

Brannan, K. M., Mostahjimi, S., McClellan, P. W. and Inandar, S. 2000. Animal waste BMP impacts on sediment and nutrient losses in runoff from the Owl Run Watershed. *Trans. ASAE* 43, 1135–66.

Breeuwsma, A. and Silva, S. 1992. Phosphorus fertilisation and environmental effects in the Netherlands and the Po region (Italy). Report 57. DLO The Winand Staring Centre, Wageningen, The Netherlands.

Butler, J. S. and Coale, F. J. 2005. Phosphorus leaching in manure-amended Atlantic Coastal Plain soils. *J. Environ. Qual.* 34(1), 370–81.

Campbell, R. E. 1965. Phosphorus fertilizer residual effects on irrigated crops in rotation. *Soil Sci. Soc. Am. Proc.* 29(1), 67–70. doi:10.2136/sssaj1965.0361599500290001000 20x.

Carpenter, S. R. 2008. Phosphorus control is critical to mitigating eutrophication. *Proc. Natl. Acad. Sci. U.S.A.* 105(32), 11039–40.

Carpenter, S. R., Caraco, N. F., Correll, D. L., Howarth, R. W., Sharpley, A. N. and Smith, V. H. 1998. Nonpoint pollution of surface waters with phosphorus and nitrogen. *Ecol. App.* 8(3), 559–68. doi:10.1890/1051-0761(1998)008[0559:NPOSWW]2.0.CO;2.

Casson, J. P., Bennett, D. R., Nolan, S. C., Olson, B. M. and Ontkean, G. R. 2006. Degree of phosphorus saturation thresholds in manure-amended soils of Alberta. *J. Environ. Qual.* 35(6), 2212–21. doi:10.2134/jeq2006.0085.

Childers, D. L., Corman, J., Edwards, M. and Elser, J. J. 2011. Sustainability challenges of phosphorus and food: solutions from closing the human phosphorus cycle. *BioScience* 61(2), 117–24. doi:10.1525/bio.2011.61.2.6.

Chow, T. L., Rees, H. W. and Daigle, J. L. 1999. Effectiveness of terraces/grassed waterway systems for soil and water conservation: a field evaluation. *J. Soil Water Conserv.* 54, 577–83.

Collins, A. L. and Walling, D. E. 2007a. Fine-grained bed sediment storage within the main channel systems of the Frome and Piddle catchments, Dorset, UK. *Hydrol. Process.* 21(11), 1448–59. doi:10.1002/hyp.6269.

Collins, A. L. and Walling, D. E. 2007b. The storage and provenance of fine sediment on the channel bed of two contrasting lowland permeable catchments, UK. *River Res. Applic.* 23(4), 429–50. doi:10.1002/rra.992.

Cox, F. R. 1994. Current phosphorus availability indices: characteristics and shortcomings. In: Havlin, J. L., Jacobsen, J. S., Lekiam, D. F., Fixen, P. E. and Hergert, G. W. (Eds), *Soil Testing: Prospects for Improving Nutrient Recommendations*. Soil Sci. Soc. Am. Special Pub. No. 40. Soil Science Society of America, Madison, WI, pp. 101–14.

Cox, F. R., Kamprath, E. J. and McCollum, R. E. 1981. A descriptive model of soil test nutrient levels following fertilization. *Soil Sci. Soc. Am. J.* 45(3), 529–32. doi:10.2136/sssaj1981.03615995004500030018x.

Daloğlu, I., Cho, K. H. and Scavia, D. 2012. Evaluating causes of trends in long-term dissolved reactive phosphorus loads to Lake Erie. *Environ. Sci. Technol.* 46(19), 10660–6. doi:10.1021/es302315d.

Dampney, P. M. R., Lord, E. I. and Chambers, B. J. 2000. Development of advice for farmers and advisors. *Soil Use Manage.* 16, 162–6.

Daverede, I. C., Kravchenko, A. N., Hoeft, R. G., Nafziger, E. D., Bullock, D. G., Warren, J. J. and Gonzini, L. C. 2003. Phosphorus runoff: effect of tillage and soil phosphorus levels. *J. Environ. Qual.* 32(4), 1436–44.

Djodjic, F., Ulen, B. and Bergström, L. 2000. Temporal and spatial variations of phosphorus losses and drainage in a structured clay soil. *Water Res.* 34(5), 1687–95. doi:10.1016/S0043-1354(99)00312-7.

Dodd, J. R. and Mallarino, A. P. 2005. Soil-test phosphorus and crop grain yield responses to long-term phosphorus fertilization for corn-soybean rotations. *Soil Sci. Soc. Am. J.* 69(4), 1118–28. doi:10.2136/sssaj2004.0279.

Dodd, R. J. and Sharpley, A. N. 2015. Recognizing the role of soil organic phosphorus in soil fertility and water quality. *Resour. Conserv. Recycl.* 105, 282–93. doi:10.1016/j.resconrec.2015.10.001.

Dodd, R. J. and Sharpley, A. N. 2016. Conservation practice effectiveness and adoption: unintended consequences and implications for sustainable phosphorus management. *Nutr. Cycl. Agroecosyst.* 104(3), 373–92. doi:10.1007/s10705-015-9748-8.

Dubrovsky, N. M., Burow, K. R., Clark, G. M., Gronberg, J. M., Hamilton, P. A., Hitt, K. J., Mueller, D. K., Munn, M. D., Nolan, B. T., Puckett, L. J., et al. 2010. The quality of our nation's waters—nutrients in the nation's streams and groundwater, 1992–2004. U.S. Geological Survey Circular 1350, 174pp. Available at: http://water.usgs.gov/nawqa/nutrients/pubs/circ1350.

Duncan, E. W., King, K. W., Williams, M. R., LaBarge, G., Pease, L. A., Smith, D. R. and Fausey, N. R. 2017. Linking soil phosphorus to dissolved phosphorus losses in the Midwest. *Agric. Environ. Lett.* 2(1), 1–5. doi:10.2134/ael2017.02.0004. Available at: https://dl.sciencesocieties.org/publications/ael/articles/2/1/170004?highlight=&search-result=1.

Edwards, D. R. and Daniel, T. C. 1993. Drying interval effects on runoff from fescue plots receiving swine manure. *Trans. ASAE* 36, 1673-8.

Elser, J. J. and Bennett, E. 2011. Phosphorus cycle: a broken biogeochemical cycle. *Nature* 478(7367), 29-31. doi:10.1038/478029a.

Evans, R., Cuffman-Neff, L. C. and Nehring, R. 1996. Increases in agricultural productivity, 1948-1993. Updates on Agricultural Resources and Environmental Indicators No. 6. US Department of Agriculture-Economic Research Service and US Govt. Printing Office, Washington DC, 85pp.

Gburek, W. J. and Sharpley, A. N. 1998. Hydrologic controls on phosphorus loss from upland agricultural watersheds. *J. Environ. Qual.* 27(2), 267-77. doi:10.2134/jeq199 8.00472425002700020005x.

Gburek, W. J., Sharpley, A. N. and Pionke, H. B. 1996. Identification of critical source areas for phosphorus export from agricultural catchments. In: Anderson, M. G. and Brookes, S. (Eds), *Advances in Hillslope Processes*. John Wiley & Sons, Chichester, England, pp. 263-82.

Halvorson, A. D. and Black, A. L. 1985. Long-term dryland crop responses to residual phosphorus fertilizer. *Soil Sci. Soc. Am. J.* 49(4), 928-33. doi:10.2136/sssaj1985.036 15995004900040028x.

Halvorson, A. D., Black, A. L., Watt, D. L. and Leholm, A. G. 1986. Economics of a one-time phosphorus application in the northern Great Plains. *Appl. Agric. Res.* 1, 137-44.

Haygarth, P. M., Chapman, P. J., Jarvis, S. C. and Smith, R. V. 1998. Phosphorus budgets for two contrasting grassland farming systems in the UK. *Soil Use Manage.* 14(s4), 160-7. doi:10.1111/j.1475-2743.1998.tb00635.x.

Haygarth, P. M., Jarvie, H. P., Powers, S. M., Sharpley, A. N., Elser, J. J., Shen, J., Peterson, H. M., Chan, N. I., Howden, N. J. K., Burt, T., et al. 2014. Sustainable phosphorus management and the need for a long-term perspective: the legacy hypothesis. *Environ. Sci. Technol.* 48(15), 8417-9. doi:10.1021/es502852s.

Heckrath, G., Brookes, P. C., Poulton, P. R. and Goulding, K. W. T. 1995. Phosphorus leaching from soils containing different P concentrations in the Broadbalk Experiment. *J. Environ. Qual.* 24, 904-10.

Hein, L. and Leemans, R. 2012. The impact of first-generation biofuels on the depletion of the global phosphorus reserve. *Ambio* 41, S341-9.

Hoffmann, C. C., Kjaergaard, C., Uusi-Kämppä, J., Hansen, H. C. and Kronvang, B. 2009. Phosphorus retention in riparian buffers: review of their efficiency. *J. Environ. Qual.* 38(5), 1942-55. doi:10.2134/jeq2008.0087.

Hooda, P. S., Rendell, A. R., Edwards, A. C., Withers, P. J. A., Aitken, M. N. and Truesdale, V. W. 2000. Relating soil phosphorus indices to potential phosphorus release to water. *J. Environ. Qual.* 29(4), 1166-71. doi:10.2134/jeq2000.00472425002900040018x.

Howarth, R. W., Anderson, D. A., Church, T. M., Greening, H., Hopkinson, C. S., Huber, W., Marcus, N., Naiman, R. J., Segerson, K., Sharpley, A. N., et al. 2000. *Clean Coastal Waters: Understanding and Reducing the Effects of Nutrient Pollution*. National Research Council. National Academy Press, Washington DC, 405pp.

Inandar, S. P., Mostaghimi, S., McClellan, P. W. and Brannan, K. M. 2001. BMP impacts on sediment and nutrient yields from an agricultural watershed in the coastal plain region. *Trans. ASAE* 44, 1191-200.

International Fertilizer Association. 2009. The Global '4R' nutrient stewardship framework: developing fertilizer best management practices for delivering economic, social and environmental benefits. International Fertilizer Industry Association (IFA), Paris,

France, 10pp. Available at: http://www.ipni.net/ipniweb/portal.nsf/0/F14DC337A C2F848B85257A8C0054E006/$FILE/The%20Global%204R%20Nutrient%20Stew ardship%20Framework.pdf.

International Plant Nutrition Institute. 2014. 4R Nutrient stewardship portal. Available at: http://www.ipni.net/4R.

Iowa Nutrient Reduction Strategy. 2012. A science and technology-based framework to assess and reduce nutrients to Iowa waters and the Gulf of Mexico, 192pp. Available at http://www.nutrientstrategy.iastate.edu/sites/default/files/documents/ NRSfull.pdf.

Jackson, C. R., Martin, J. K., Leigh, D. S. and West, L. T. 2005. A southeastern Piedmont watershed sediment budget: evidence for a multi-millennial agricultural legacy. *J. Soil Water Conserv.* 60, 298-310.

Jarvie, H. P., Jurgens, M. D., Williams, R. J., Neal, C., Davies, J. J. L., Barrett, C. and White, J. 2005. Role of river bed sediments as sources and sinks of phosphorus across two major eutrophic UK river basins: the Hampshire Avon and Herefordshire Wye. *J. Hydrol.* 304(1-4), 51-74. doi:10.1016/j.jhydrol.2004.10.002.

Jarvie, H. P., Sharpley, A. N., Spears, B., Buda, A. R., May, L. and Kleinman, P. J. A. 2013. Water quality remediation faces unprecedented challenges from 'legacy phosphorus'. *Environ. Sci. Technol.* 47(16), 8997-8. doi:10.1021/es403160a.

Jarvie, H. P., Sharpley, A. N., Flaten, D., Kleinman, P. J. A., Jenkins, A. and Simmons, T. 2015. The pivotal role of phosphorus in a resilient water-energy-food security nexus. *J. Environ. Qual.* 44(5), 1308-26.

Jarvie, H. P., Johnson, L. T., Sharpley, A. N., Smith, D. R., Baker, D. B., Bruulsema, T. W. and Confesor, R. 2017. Increased soluble phosphorus loads to Lake Erie: unintended consequences of conservation practices? *J. Environ. Qual.* 46(1), 123-32. doi:10.2134/jeq2016.07.0248.

Jasinksi, S. 2015. Phosphate rock. In: *Mineral Commodity Summaries 2014*. U.S. Department of the Interior and U.S. Geological Survey. Available at: http://minerals .usgs.gov/minerals/pubs/mcs/2014/mcs2014.pdf.

Jordan, T. E., Whigham, D. F., Hofmockel, K. H. and Pittek, M. A. 2003. Nutrient and sediment removal by a restored wetland receiving agricultural runoff. *J. Environ. Qual.* 32(4), 1534-47. doi:10.2134/jeq2003.1534.

Jose, H. D. 1981. An economic comparison of batch and annual phosphorus fertilizer application in wheat production in western Canada. *Can. J. Soil Sci.* 61(1), 47-54. doi:10.4141/cjss81-006.

Khan, S., Khan, M. A., Hanjra, M. A. and Mu, J. 2009. Pathways to reduce the environmental footprints of water and energy inputs in food production. *Food Policy* 34(2), 141-9. doi:10.1016/j.foodpol.2008.11.002.

King, K. W., Williams, M. R., LaBarge, G. A., Smith, D. R., Reutter, J. M., Duncan, E. W. and Pease, L. A. 2018. Addressing agricultural phosphorus loss in artificially drained landscapes with 4R nutrient management practices. *J. Soil Water Conserv.* 73(1), 35-47. doi:10.2489/jswc.73.1.35.

Kleinman, P. J. A. and Sharpley, A. N. 2002. Estimating soil phosphorus sorption saturation from Mehlich-3 data. *Commun. Soil Sci. Plant Anal.* 33(11-12), 1825-39. doi:10.1081/ CSS-120004825.

Kleinman, P. J. A., Salon, P., Sharpley, A. N. and Saporito, L. S. 2005. Effect of cover crops established at time of corn planting on phosphorus runoff from soils before and after dairy manure application. *J. Soil Water Conserv.* 60, 311-22.

Kleinman, P. J. A., Sharpley, A. N., Withers, P. J. A., Bergström, L., Johnson, L. T. and Doody, D. G. 2015. Implementing agricultural phosphorus science and management to combat eutrophication. *Ambio* 44(Suppl. 2), S297–310.

Kleinman, P. J. A., Sharpley, A. N., Buda, A. R., Easton, Z. M., Lory, J. A., Osmond, D. L., Radcliffe, D. E., Nelson, N. O., Veith, T. L. and Doody, D. G. 2017. The promise, practice, and state of planning tools to assess site vulnerability to runoff phosphorus loss. *J. Environ. Qual.* 46(6), 1243–9. doi:10.2134/jeq2017.10.0395.

Kohler, J., Hilt, S., Adrian, R., Nicklisch, A., Kozerski, H. P. and Walz, N. 2005. Long-term response of a shallow, moderately flushed lake to reduced external phosphorus and nitrogen loading. *Freshw. Biol.* 50(10), 1639–50. doi:10.1111/j.1365-2427.2005.01430.x.

Kovzelove, C., Simpson, T. and Korcak, R. 2010. Quantification and implications of surplus phosphorus and manure in major animal production. Regions of Maryland, Pennsylvania and Virginia. Water Stewardship, Annapolis, MD, 56pp. Available at http://waterstewardshipinc.org/downloads/P_PAPER_FINAL_2-9-10.pdf.

Kröger, R., Perez, M., Walker, S. and Sharpley, A. N. 2012. Review of best management practice reduction efficiencies in the Lower Mississippi Alluvial Valley. *J. Soil Water Conserv.* 67(6), 556–63. doi:10.2489/jswc.67.6.556.

Kronvang, B., Bechmann, M., Lundekvam, H., Behrendt, H., Rubaek, G. H., Schoumans, O. F., Syversen, N., Andersen, H. E. and Hoffmann, C. C. 2005. Phosphorus losses from agricultural areas in river basins: effects and uncertainties of targeted mitigation measures. *J. Environ. Qual.* 34(6), 2129–44. doi:10.2134/jeq2004.0439.

Lamond, R. E. 1987. Comparison of fertilizer solution placement methods for grain sorghum under two tillage systems. *J. Fert. Issues* 4, 43–7.

Lander, C. H., Moffitt, D. and Alt, K. 1998. Nutrients available from livestock manure relative to crop growth requirements. Resource Assessment and Strategic Planning Working Paper 98-1. USDA-Natural Resources Conservation Service, Washington DC. Available at: http://www.nrcs.usda.gov/technical/land/pubs/nlweb.html (Last verified 1 September 2005).

Lanyon, L. E. 2005. Phosphorus, animal nutrition and feeding: overview. In: Sims, J. T. and Sharpley, A. N. (Eds), *Phosphorus; Agriculture and the Environment*. Am. Soc. Agron. Monograph. American Society of Agronomy, Madison, WI, pp. 561–86.

Larsen, D. P., Sickle, J. V., Malueg, K. W. and Smith, P. D. 1979. The effect of wastewater phosphorus removal on Shagawa Lake, Minnesota: phosphorus supplies, lake phosphorus and chlorophyll a. *Water Res.* 13(12), 1259–72. doi:10.1016/0043-1354(79)90170-2.

Leinweber, P., Lunsmann, F. and Eckhardt, K. U. 1997. Phosphorus sorption capacities and saturation of soils in two regions with different livestock densities in northwest Germany. *Soil Use Manage.* 13(2), 82–9. doi:10.1111/j.1475-2743.1997. tb00562.x.

Li, H., Liu, J., Li, G., Shen, J., Bergström, L. and Zhang, F. 2015. Past, present and future use of phosphorus in Chinese agriculture and its influence on phosphorus losses. *Ambio* 44, S274–85.

Lookman, R., Jansen, K., Merckx, R. and Vlassak, K. 1996. Relationship between soil properties and phosphate saturation parameters: a transect study in northern Belgium. *Geoderma* 69(3–4), 265–74. doi:10.1016/0016-7061(95)00068-2.

Maguire, R. O. and Sims, J. T. 2002a. Soil testing to predict phosphorus leaching. *J. Environ. Qual.* 31(5), 1601–9. doi:10.2134/jeq2002.1601.

Maguire, R. O. and Sims, J. T. 2002b. Measuring agronomic and environmental soil phosphorus saturation and predicting phosphorus leaching with Mehlich 3. *Soil Sci. Soc. Am. J.* 66(6), 2033-9. doi:10.2136/sssaj2002.2033.

McCollum, R. E. 1991. Buildup and decline in soil phosphorus: 30-year trends on a Typic Umprabuult. *Agron. J.* 83(1), 77-85. doi:10.2134/agronj1991.00021962008300010019x.

McDowell, R. W. and Sharpley, A. N. 2001. Approximating phosphorus release from soils to surface runoff and subsurface drainage. *J. Environ. Qual.* 30(2), 508-20. doi:10.2134/jeq2001.302508x.

Meals, D. W. and Hopkins, R. B. 2002. Phosphorus reductions following riparian restoration in two agricultural watersheds in Vermont, USA. *Water Sci. Technol.* 45(9), 51-60. doi:10.2166/wst.2002.0203.

Meals, D. W., Dressing, S. A. and Davenport, T. E. 2010. Lag time in water quality response to best management practices: a review. *J. Environ. Qual.* 39(1), 85-96. Review. doi:10.2134/jeq2009.0108.

Michalak, A. M., Anderson, E. J., Beletsky, D., Boland, S., Bosch, N. S., Bridgeman, T. B., Chaffin, J. D., Cho, K., Confesor, R., Daloğlu, I., et al. 2013. Record-setting algal bloom in Lake Erie caused by agricultural and meteorological trends consistent with expected future conditions. *Proc. Natl. Acad. Sci. U.S.A.* 110(16), 6448-52. doi:10.1073/pnas.1216006110.

Miltner, R. J. 2015. Measuring the contribution of agricultural conservation practices to observed trends and recent condition in water quality indicators in Ohio, USA. *J. Environ. Qual.* 44(6), 1821-31. doi:10.2134/jeq2014.12.0550.

Moss, B., Barker, T., Stephen, D., Williams, A. E., Balayla, D. J., Beklioglu, M. and Carvalhol, L. 2005. Consequences of reduced nutrient loading on a lake system in a lowland catchment: deviation from the norm? *Freshw. Biol.* 50(10), 1687-705. doi:10.1111/j.1365-2427.2005.01416.x.

Mueller, D. H., Wendt, R. C. and Daniel, T. C. 1984. Phosphorus losses as affected by tillage and manure application. *Soil Sci. Soc. Am. J.* 48(4), 901-5. doi:10.2136/sssaj1984.03615995004800040040x.

Murphy, P. N. C., Mellander, P. -E., Melland, A. R., Buckley, C., Shore, M., Shortle, G., Wall, D. P., Treacy, M., Shine, O., Mechan, S., et al. 2015. Variable response to phosphorus mitigation measures across the nutrient transfer continuum in a dairy grassland catchment. *Agric. Ecosyst. Environ.* 207, 192-202. doi:10.1016/j.agee.2015.04.008.

Nair, V. D. and Harris, W. G. 2004. A capacity factor as an alternative to soil test phosphorus in phosphorus risk assessment. *N. Z. J. Agric. Res.* 47(4), 491-7. doi:10.1080/00288233.2004.9513616.

Nair, V. D. and Harris, W. G. 2014. Soil phosphorus storage capacity for environmental risk assessment. *Adv. Agric.* 2014, Article ID 723064, 9pp. doi:10.1155/2014/723064.

National Research Council. 2008. Nutrient control actions for improving water quality in the Mississippi River Basin and northern Gulf of Mexico. Committee on the Mississippi River and the Clean Water Act: scientific, Modeling and Technical Aspects of Nutrient Pollutant Load Allocation and Implementation. National Research Council, Washington DC, 75pp. Available at: http://www.nap.edu/catalog/12544.html.

Nelson, N. O., Parsons, J. E. and Mikkelsen, R. L. 2005. Field-scale evaluation of phosphorus leaching in acid sandy soils receiving swine waste. *J. Environ. Qual.* 34(6), 2024-35. doi:10.2134/jeq2004.0445.

Neset, T.-S. S. and Cordell, D. 2012. Global phosphorus scarcity: identifying synergies for a sustainable future. *J. Sci. Food Agric.* 92(1), 2-6. doi:10.1002/jsfa.4650.

Osmond, D., Meals, D., Hoag, D., Arabi, M., Luloff, A., Jennings, G., McFarland, M., Spooner, J., Sharpley, A. N. and Line, D. 2012. Improving conservation practices programming to protect water quality in agricultural watersheds: lessons learned from the National Institute of Food and Agriculture Conservation Effects Assessment Project. *J. Soil Water Conserv.* 67(5), 122A-7A. doi:10.2489/jswc.67.5.122A.

Owens, P. N. and Walling, D. E. 2002. The phosphorus content of fluvial sediment in rural and industrialized river basins. *Water Res.* 36(3), 685-701. doi:10.1016/S0043-1354(01)00247-0.

Phillips, G., Kelly, A., Pitt, J. A., Sanderson, R. and Taylor, E. 2005. The recovery of a very shallow eutrophic lake, 20 years after the control of effluent derived phosphorus. *Freshw. Biol.* 50(10), 1628-38. doi:10.1111/j.1365-2427.2005.01434.x.

Pierson, S. T., Cabrera, M. L., Evanylo, G. K., Kuykendall, H. A., Hoveland, C. S., McCann, M. A. and West, L. T. 2001. Phosphorus and ammonium concentrations in surface runoff from grasslands fertilized with broiler litter. *J. Environ. Qual.* 30(5), 1784-9. doi:10.2134/jeq2001.3051784x.

Pierzynski, G. M. and Logan, T. J. 1993. Crop, soil, and management effects on phosphorus soil test levels. *J. Prod. Agric.* 6, 513-20.

Pionke, H. B., Gburek, W. J., Sharpley, A. N. and Schnabel, R. R. 1996. Flow and nutrient export patterns for an agricultural hill-land watershed. *Water Resour. Res.* 32(6), 1795-804. doi:10.1029/96WR00637.

Pionke, H. B., Gburek, W. J., Schnabel, R. R., Sharpley, A. N. and Elwinger, G. F. 1999. Seasonal flow, nutrient concentrations and loading patterns in stream flow draining an agricultural hill-land watershed. *J. Hydrol.* 220(1-2), 62-73. doi:10.1016/S0022-1694(99)00064-5.

Pionke, H. B., Gburek, W. J. and Sharpley, A. N. 2000. Critical source area controls on water quality in an agricultural watershed located in the Chesapeake Basin. *Ecol. Eng.* 14(4), 325-35. doi:10.1016/S0925-8574(99)00059-2.

Pote, D. H., Daniel, T. C., Moore, P. A., Nichols, D. J., Sharpley, A. N. and Edwards, D. R. 1996. Relating extractable soil phosphorus to phosphorus losses in runoff. *Soil Sci. Soc. Am. J.* 60(3), 855-9. doi:10.2136/sssaj1996.03615995006000030025x.

Reimer, A. P., Thompson, A. W. and Prokopy, L. S. 2012. The multi-dimensional nature of environmental attitudes among farmers in Indiana: implications for conservation adoption. *Agric. Hum. Values* 29(1), 29-40. doi:10.1007/s10460-011-9308-z.

Richards, R. P., Baker, D. B. and Eckert, D. J. 2002. Trends in agriculture in the LEASEQ watersheds, 1975-1995. Lake Erie Agricultural Systems for Environmental Quality. *J. Environ. Qual.* 31(1), 17-24.

Richards, R. P., Baker, D. B. and Crumrine, J. P. 2009. Improved water quality in Ohio tributaries to Lake Erie: A consequence of conservation practices. *J. Soil Water Conserv.* 64(3), 200-11. doi:10.2489/jswc.64.3.200.

Richards, R. P., Baker, D. B., Crumrine, J. P. and Sterns, A. M. 2010. Unusually large loads in 2007 from the Maumee and Sandusky Rivers, tributaries to Lake Erie. *J. Soil Water Conserv.* 65(6), 450-62. doi:10.2489/jswc.65.6.450.

Robertson, G. P., Dale, V. H., Doering, O. C., Hamburg, S. P., Melillo, J. M., Wander, M. M., Parton, W. J., Adler, P. R., Barney, J. N., Cruse, R. M., et al. 2008. Agriculture - Sustainable biofuels Redux. *Science* 322(5898), 49-50. doi:10.1126/science.1161525.

Schindler, D. W., Hecky, R. E., Findlay, D. L., Stainton, M. P., Parker, B. R., Paterson, M. J., Beaty, K. G., Lyng, M. and Kasian, S. E. M. 2008. Eutrophication of lakes cannot be controlled by reducing nitrogen input: results of a 37-year whole-ecosystem experiment. *Proc. Natl. Acad. Sci. U.S.A.* 105(32), 11254–8. doi:10.1073/pnas.0805108105. Available at: http://www.pnas.org/content/105/32/11254.full.pdf.

Scholz, R. W. and Wellmer, F. W. 2013. Approaching a dynamic view on the availability of mineral resources: what we may learn from the case of phosphorus? *Global Environ. Chang.* 23, 11–27. doi:10.1016/j.gloenvcha.2012.10.013.

Schönbrunner, I. M., Preiner, S. and Hein, T. 2012. Impact of drying and re-flooding of sediment on phosphorus dynamics of river-floodplain systems. *Sci. Total Environ.* 432, 329–37. doi:10.1016/j.scitotenv.2012.06.025.

Schoumans, O. F. and Breeuwsma, A. 1997. The relation between accumulation and leaching of phosphorus: laboratory, field and modelling results. In: Tunney, H., Carton, O. T., Brookes, P. C. and Johnston, A. E. (Eds), *Phosphorus Loss from Soil to Water*. CAB International Press, Cambridge, England, pp. 361–3.

Schoumans, O. F., Bouraoui, F., Kabbe, C., Oenema, O. and van Dijk, K. C. 2015. Phosphorus management in Europe in a changing world. *Ambio* 44(Suppl. 2), S180–92. doi:10.1007/s13280-014-0613-9.

Schroeder, P. D., Radcliffe, D. E. and Cabrera, M. L. 2004. Rainfall timing and poultry litter application rate effects on phosphorus loss in surface runoff. *J. Environ. Qual.* 33(6), 2201–9. doi:10.2134/jeq2004.2201.

Sharpley, A. N. 1995a. Identifying sites vulnerable to phosphorus loss in agricultural runoff. *J. Environ. Qual.* 24(5), 947–51. doi:10.2134/jeq1995.00472425002400050024x.

Sharpley, A. N. 1995b. Dependence of runoff phosphorus on soil phosphorus. *J. Environ. Qual.* 24, 920–6.

Sharpley, A. N. 1997. Rainfall frequency and nitrogen and phosphorus in runoff from soil amended with poultry litter. *J. Environ. Qual.* 26, 1127–32.

Sharpley, A. 2018. Agriculture, nutrient management and water quality. *Reference Module in Life Sciences*. Elsevier. Available at: https://www.sciencedirect.com/science/article/pii/B9780128096338207589.

Sharpley, A. N. and Smith, S. J. 1991. Effect of cover crops on surface water quality. In: Hargrove, W. L. (Ed.), *Cover Crops for Clean Water*. Soil and Water Conservation Society, Ankeny, IA, pp. 41–50.

Sharpley, A. N. and Smith, S. J. 1994. Wheat tillage and water quality in the Southern Plains. *Soil Till. Res.* 30(1), 33–48. doi:10.1016/0167-1987(94)90149-X.

Sharpley, A. N. and Syers, J. K. 1979. Loss of nitrogen and phosphorus in tile drainage as influenced by urea application and grazing animals. *N. Z. J. Agric. Res.* 22(1), 127–31. doi:10.1080/00288233.1979.10420852.

Sharpley, A. N. and Tunney, H. 2000. Phosphorus research strategies to meet agricultural and environmental challenges of the 21st century. *J. Environ. Qual.* 29(1), 176–81. doi:10.2134/jeq2000.00472425002900010022x.

Sharpley, A. N., McDowell, R. W. and Kleinman, P. J. A. 2001. Phosphorus loss from land to water: integrating agricultural and environmental management. *Plant Soil* 237(2), 287–307. doi:10.1023/A:1013335814593.

Sharpley, A. N., Weld, J. L., Beegle, D. B., Kleinman, P. J. A., Gburek, W. J., Moore, P. A. and Mullins, G. 2003. Development of phosphorus indices for nutrient management planning strategies in the U.S. *J. Soil Water Conserv.* 58, 137–52.

Sharpley, A. N., Herron, S. and Daniel, T. C. 2007. Overcoming the challenges of phosphorus-based management in poultry farming. *J. Soil Water Conserv.* 62, 375–89.

Sharpley, A. N., Kleinman, P. J. A., Jordan, P., Bergström, L. and Allen, A. L. 2009. Evaluating the success of phosphorus management from field to watershed. *J. Environ. Qual.* 38(5), 1981–8. doi:10.2134/jeq2008.0056.

Sharpley, A. N., Beegle, D. G., Bolster, C., Good, L. W., Joern, B., Ketterings, Q., Lory, J., Mikkelsen, R., Osmond, D. and Vadas, P. A. 2012. Phosphorus indices: why we need to take stock of how we are doing. *J. Environ. Qual.* 41(6), 1711–9. doi:10.2134/jeq2012.0040.

Sharpley, A. N., Jarvie, H. P., Buda, A., May, L., Spears, B. and Kleinman, P. 2013. Phosphorus legacy: overcoming the effects of past management practices to mitigate future water quality impairment. *J. Environ. Qual.* 42(5), 1308–26. doi:10.2134/jeq2013.03.0098.

Sharpley, A. N., Kleinman, P. J. A., Baffaut, C., Beegle, D., Bolster, C., Collick, A., Easton, Z., Lory, J., Nelson, N., Osmond, D., et al. 2017. Evaluation of phosphorus site assessment tools: lessons from the U.S.A. *J. Environ. Qual.* 46(6), 1250–6. doi:10.2134/jeq2016.11.0427.

Shulte, R. P. O., Melland, A. R., Fenton, O., Herlhy, M., Richards, K. and Jordan, P. 2010. Modelling soil phosphorus decline: expectations of Water Framework Directive policies. *Environ. Sci. Policy* 13(6), 472–84. doi:10.1016/j.envsci.2010.06.002.

Sims, J. T. and Kleinman, P. J. A. 2005. Managing agricultural phosphorus for environmental protection. In: Sims, J. T. and Sharpley, A. N. (Eds), *Phosphorus; Agriculture and the Environment*. Am. Soc. Agron. Monograph. American Society of Agronomy, Madison, WI, pp. 1021–68.

Sims, J. T. and Sharpley, A. N. (Ed.). 2005. *Phosphorus: Agriculture and the Environment*. Am. Soc. Agron. Monograph. American Society of Agronomy, Madison, WI.

Sims, J. T., Joern, B. C. and Simard, R. R. 1998. Phosphorus losses in agricultural drainage: historical perspective and current research. *J. Environ. Qual.* 27, 277–93.

Sims, J. T., Maguire, R. O., Leytem, A. B., Gartley, K. L. and Pautler, M. C. 2002. Evaluation of Mehlich 3 as an agri-environmental soil phosphorus test for the Mid-Atlantic United States of America. *Soil Sci. Soc. Am. J.* 66(6), 2016–32. doi:10.2136/sssaj2002.2016.

Smith, D. R., Owens, P. R., Leytem, A. B. and Warnemuende, E. A. 2007. Nutrient losses from manure and fertilizer applications as impacted by time to first runoff event. *Environ. Pollut.* 147(1), 131–7. doi:10.1016/j.envpol.2006.08.021.

Smith, D. R., King, K. W. and Williams, M. R. 2015a. What is causing the harmful algal blooms in Lake Erie? *J. Soil Water Conserv.* 70(2), 27A–9A. doi:10.2489/jswc.70.2.27A.

Smith, D. R., King, K. W., Johnson, L., Francesconi, W., Richards, P., Baker, D. and Sharpley, A. N. 2015b. Surface runoff and tile drainage transport of phosphorus in the Midwestern United States. *J. Environ. Qual.* 44(2), 495–502. doi:10.2134/jeq2014.04.0176.

Smith, D. R., Wilson, R. S., King, K. W., Zwonitzer, M., McGrath, J. M., Harmel, R. D., Haney, R. L. and Johnson, L. T. 2018. Lake Erie, phosphorus, and microcystin: is it really the farmer's fault? *J. Soil Water Conserv.* 73(1), 48–57. doi:10.2489/jswc.73.1.48.

Spratt, E. D., Warder, F. G., Bailey, L. D. and Read, D. W. L. 1980. Measurement of fertilizer phosphorus residues and its utilization. *Soil Sci. Soc. Am. J.* 44(6), 1200–4. doi:10.2136/sssaj1980.03615995004400060013x.

Stumpf, R. P., Johnson, L. T., Wynne, T. T. and Baker, D. B. 2016. Forecasting annual cyanobacterial bloom biomass to inform management decisions in Lake Erie. *J. Great Lakes Res.* 42(6), 1174–83, doi:10.1016/j.jglr.2016.08.006.

Sun, B., Zhang, L., Yang, L., Zhang, F., Norse, D. and Zhu, Z. 2012. Agricultural non-point source pollution in China: causes and mitigation measures. *Ambio* 41(4), S370–9. doi:10.1007/s13280-012-0249-6.

Surridge, B. W. J., Heathwaite, A. L. and Baird, A. J. 2012. Phosphorus mobilisation and transport within a long-restored floodplain wetland. *Ecol. Eng.* 44, 348–59. doi:10.1016/j.ecoleng.2012.02.009.

Tiessen, K. H. D., Elliot, J. A., Yarotski, J., Lobb, D. A., Flaton, D. N. and Glozier, N. E. 2010. Conventional and conservation tillage: influence on seasonal runoff, sediment, and nutrient losses in the Canadian Prairies. *J. Environ. Qual.* 39(3), 964–80. doi:10.2134/jeq2009.0219.

Tilman, D., Socolow, R., Foley, J. A., Hill, J., Larson, E., Lynd, L., Pacala, S., Reilly, J., Searchinger, T., Somerville, C., et al. 2009. Beneficial biofuels-the food, energy, and environment trilemma. *Science* 325(5938), 270–1. doi:10.1126/science.1177970.

Torbert, H. A., Daniel, T. C., Lemunyon, J. L. and Jones, R. M. 2002. Relationship of soil test phosphorus and sampling depth to runoff phosphorus in calcareous and noncalcareous soils. *J. Environ. Qual.* 31(4), 1380–7. doi:10.2134/jeq2002.1380.

Trimble, S. W. 2010. Streams, valleys and floodplains in the sediment cascade. In: Burt, T. and Allison, R. (Eds), *Sediment Cascades: an Integrated Approach*. John Wiley & Sons, Ltd. ISBN: 978-0-470-84962-0.

U.S. Department of Agriculture and U.S. Environmental Protection Agency. 1999. Unified national strategy for Animal Feeding Operations. 9 March 1999. Available at: http://www.epa.gov/npdes/pubs/finafost.pdf.

U.S. Environmental Protection Agency. 2010. Guidance for Federal Land Management in the Chesapeake Bay Watershed. Chapter 2: Agriculture. EPA841-R-10-002. Office of Wetlands, Oceans, and Watersheds and U.S. Environmental Protection Agency, Washington DC, 247pp. Available at: http://www.epa.gov/owow:keep/NPS/chesbay502/pdf/chesbay_chap02.pdf.

Uusi-Kämppä, J., Braskerud, B., Jansson, H., Syverson, N. and Uusitalo, R. 2000. Buffer zones and constructed wetlands as filters for agricultural phosphorus. *J. Environ. Qual.* 29(1), 151–8. doi:10.2134/jeq2000.00472425002900010019x.

Walling, D. E., Owens, P. N., Carter, J., Leeks, G. J. L., Lewis, S., Meharg, A. A. and Wright, J. 2003. Storage of sediment-associated nutrients and contaminants in river channel and floodplain systems. *Appl. Geochem.* 18(2), 195–220. doi:10.1016/S0883-2927(02)00121-X.

Welch, L. F., Mulvaney, D. L., Boone, L. V., McKibben, G. E. and Pendleton, J. W. 1966. Relative efficiency of broadcast versus banded phosphorus for corn. *Agron. J.* 58(3), 283–7. doi:10.2134/agronj1966.00021962005800030011x.

Westerman, P. W., Donnely, T. L. and Overcash, M. R. 1983. Erosion of soil and poultry manure – a laboratory study. *Trans. ASAE* 26, 1070–8, 1084.

Williams, M. R., King, K. W., Ford, W., Buda, A. R. and Kennedy, C. D. 2016. Effect of tillage on macropore flow and phosphorus transport to tile drains. *Water Resour. Res.* 52, 2868–82. doi:10.1002/2015WR017650.

Wines, M. 2014. Behind Toledo's water crisis, a long-troubled lake Erie. *New York Times*. 4 August. Available at: https://www.nytimes.com/2014/08/05/us/lifting-ban-toledo-says-its-water-is-safe-to-drink-again.html?_r=0.

Withers, P. J. A. and Jarvie, H. P. 2008. Delivery and cycling of phosphorus in UK rivers: implications for catchment management. *Sci. Total Environ.* 400(1-3), 379–95. doi:10.1016/j.scitotenv.2008.08.002.

Withers, P. J. A., Edwards, A. C. and Foy, R. H. 2002. Phosphorus cycling in UK agriculture and implications for phosphorus loss from soil. *Soil Use Manage.* 17, 139–49.

Withers, P. J. A., Sylvester-Bradley, R., Jones, D. L., Healey, J. R. and Talboys, P. J. 2014. Feed the crop not the soil: rethinking phosphorus management in the food chain. *Environ. Sci. Technol.* 48(12), 6523–30. doi:10.1021/es501670j.

Yost, R. S., Kamprath, E. J., Naderman, G. C. and Lobato, E. 1981. Residual effects of phosphorus applications on a high phosphorus adsorbing oxisol of central Brazil. *Soil Sci. Soc. Am. J.* 45(3), 540–3. doi:10.2136/sssaj1981.03615995004500030021x.

Zhang, T. Q., MacKenzie, A. F., Liang, B. C. and Drury, C. F. 2004. Soil test phosphorus and phosphorus fractions with long-term phosphorus addition and depletion. *Soil Sci. Soc. Am. J.* 68(2), 519–28. doi:10.2136/sssaj2004.5190.

CPSIA information can be obtained
at www.ICGtesting.com
Printed in the USA
JSHW011428100423
40140JS00005B/91